AT SAINT LOUIS UNIVERSITY SCHOOL OF MEDICINE

by THOMAS C. WESTFALL

Copyright © 2020, Dr. Thomas C. Westfall
All rights reserved.

No part of this publication may be reproduced or transmitted in any form or by any means, electronic or mechanical, including photocopy, recording, or any information storage and retrieval system, without permission in writing from Dr. Thomas C. Westfall.

Permissions may be sought directly from Dr. Thomas C. Westfall.

Library of Congress Control Number: 2020916105

ISBN: 9780578752440

Printed in the United States of America
20 21 22 23 24   5 4 3 2 1

Table of Contents

Preface vii
Acknowledgments viii
List of Tables ix
List of Figures xi

CHAPTER ONE
The First School of Medicine at Saint Louis University (1842–55) 1

Introduction and Background 1

William Beaumont, MD (1785–1853) 2

Medical Education Continues in St. Louis 3

Departments of Physiology and Materia Medica (Pharmacology) 4
   Early Chairs: Department of Physiology 4
   Synopsis of the Course in Physiology as Taught by Professor Holmes 7
   Early Chairs: Department of Pharmacology (Materia Medica) 7
   Synopsis of Pharmacology Course from Lectures of Professor Reyburn 8

Separation of Saint Louis Medical College from Saint Louis University 9

CHAPTER TWO
Formation of the Current School of Medicine at Saint Louis University (1903–Present) 11

CHAPTER THREE
Department of Pharmacology and Physiology: The Shattinger, Lyon, and Joseph Eras (1903–21) 15

The Shattinger Era (1903–04) 15
   Biographical Sketch of D. Shattinger, MD 15
   Members of the Department 15
   Course in Physiology 15

The Lyon Era (1904–13) 15
   Biographical Sketch of Elias Potter Lyon, PhD . . . 15
   Aims of the Department during the Lyon Era 16
   Synopsis of the Course 16
   Faculty in the Department during the Lyon Era . . 19

The Joseph Era (1913–28) 19
   Biographical Sketch of Don R. Joseph, MD 19
   Aims of the Department during the Joseph Era . . . 20
   Faculty in the Department of Physiology and Pharmacology during the Joseph Era 20

Faculty during the Lyon and Joseph Eras 20

Graduates during the Joseph Era 22

CHAPTER FOUR
The John Auer Era, Department of Pharmacology (1921–48) 23

Biographical Sketch of John Auer, MD 23

The New Department: Objectives and Aims . . . 24

Synopsis of the Pharmacology Course 25

Formation of the American Society
for Pharmacology and Experimental
Therapeutics (ASPET) in 1908............26

Fifteenth Annual Meeting of ASPET
Comes to Saint Louis University in 1923......27

The John Auer Lectureship 27

Faculty during the Auer Era 28

CHAPTER FIVE
Department of Pharmacology: The Robert H. K. Foster, Erwin Nelson, and Elijah Adams Eras (1948–63) 29

The Robert H. K. Foster Era (1948–52).......29

Biographical Sketch of Robert H. K. Foster, MD, PhD 29

Synopsis of the Pharmacology Course 29

Faculty during the Foster Era 30

The Erwin E. Nelson Era (1952–57)..........31

Biographical Sketch of Erwin E. Nelson, PhD 31

The Role of Pharmacology in Medical Education.. 31

Faculty Research Interests 32

Graduates during the Nelson Era 32

Faculty during the Nelson Era 34

The Elijah Adams Era (1958–63) 34

Biographical Sketch of Elijah Adams, PhD 34

Synopsis of the Pharmacology Course 35

Faculty during the Adams Era 36

CHAPTER SIX
Department of Pharmacology: The René Wégria Era (1963–79) 39

Biographical Sketch of René William Edward Wégria, MD, MedScD..............39

Medical and Graduate Education...........40

Faculty Research Interests..................40

Graduates during the Wégria Era...........41

Faculty during the Wégria Era 41

CHAPTER SEVEN
Department of Physiology: The Alrick B. Hertzman Era (1928–66)...............43

Biographical Sketch of Alrick B. Hertzman, PhD........................43

Teaching Activities during the Hertzman Era ..44

Research during the Hertzman Era 45

Medical Horizons: An ABC National Television Series........................45

Graduate Training and Degrees 46

Faculty during the Hertzman Era............47

Faculty Members that Went On to Leadership Roles.......................51

CHAPTER EIGHT
Department of Physiology: The Alexander Lind Era (1968–88) 53

Biographical Sketch of Alexander Lind, DPhil, DSc 53

Mission and Objectives of the Department 53

Medical and Graduate Education............54

Graduates during the Lind Era..............56

Faculty Research Interests..................56

Faculty during the Lind Era 58

CHAPTER NINE Department of Pharmacology: The Thomas C. Westfall Era, Part I (1979–90) 63

Biographical Sketch of Thomas C. Westfall, PhD 63

Recruitment of Thomas C. Westfall 65

Early Recruitments of Faculty 65

Mission and Objectives of the Department from 1979 to 1990 66

Medical Pharmacology Course, 1979 to 1998...67

Objectives of the Medical Pharmacology Course (1979-98)......................... 68

Graduate Program and Graduate Education . . . 68

   Representative Graduate Courses (1979–98). 73

Faculty Research Interests. 74

Faculty during the Westfall Era, Part I 78

CHAPTER TEN
The Department of Pharmacological and Physiological Science: The Thomas C. Westfall Era, Part II (1990–2013) 83

Faculty Who Carried Over to the New Department. 85

Recruitments to the New Department and Additions from Anatomy/Neuroscience. 85

Mission and Objectives of the Department 85

Organization and Governance of the Department. 86

Faculty Development Efforts 88

Medical Education Activities 89

Graduate Training Activities. 96

   Graduate Program and Graduate Education 96

   Development of the Core Graduate Program in Biomedical Sciences (1998–Present) 96

   Current Components of the Graduate Program. . 102

   Service Course Activities 106

   Drugs We Use and Abuse: BLA245 107

Faculty Research Interests. 107

Extramural Support 120

Formal Departmental Reviews 120

   Departmental Review, 1990-95 120

   Departmental Review, 1997–2002 122

Seminars/Visiting Speakers Series (1979–2013) 131

   Nobel Laureates 131

   Louis D'Agrosa Memorial Lecture 134

   Distinguished Alumni Lectureship 135

Faculty Meetings and Retreats 135

Social Activities 137

The Thomas C. Westfall Graduate Student Symposium 141

Faculty during the Westfall Era (1979–2013). . 144

   Tenured Faculty, Department of Pharmacology (1979–90) 145

   Tenured Faculty, Department of Pharmacological and Physiological Science (1990–2013). 145

   Secondary, Research, and Adjunct Appointments during Dr. Westfall's Tenure as Chair 146

Faculty during the Westfall Era, Part II 146

PhD Graduates during the Westfall Era (1979–2013) 155

Office Staff during the Westfall Era 156

Research Assistants and Technicians. 158

CHAPTER ELEVEN
Department of Pharmacology and Physiology: The Thomas P. Burris Era and the Mark Voigt Era (2013–18 and 2018–19). 165

Biographical Sketch of Thomas Burris, PhD . . 165

Recruitment of Dr. Thomas Burris. 166

Dr. Burris's Mission and Vision for the Department of Pharmacology and Physiology. . 167

   Education and Training. 167

   Research 167

Departmental Administration and Committee Structure. 167

Faculty Meetings and Retreats 168

Medical Education Activities 168

Principles of Pharmacology 172

Graduate Program and Educational Activities 172

   Graduate Courses. 172

Grant-Writing Course *172*
Service Courses (2013–18) *179*
Faculty Research Interests *179*
Extramural Support. *181*
Faculty during the Burris and Voigt Eras *181*

Faculty during the Burris and Voigt Eras 181

PhD Graduates and Students during the Burris and Voigt Eras *185*

Transition from the Burris Era to the Voigt Era and the Future of the Department. *185*

Faculty Who Transitioned from the Department under Dr. Burris. 186

Faculty Meetings and Retreats *187*
Update of Departmental Committees. *187*
Proposal for Faculty Evaluation *187*
Teaching Activities. *190*

Epilogue and the Future of the Department of Pharmacology and Physiology 190

Department of Pharmacology and Physiology Statistics. *193*

Closing Comments 195

References 196

Preface

The idea for this book on the history of pharmacology and physiology at Saint Louis University School of Medicine originated out of the curiosity of the author, Thomas C. Westfall, about the men and women who preceded him as chairs and faculty members in these departments of the university, starting with their inception in 1837. It was also enkindled by the fact that similar histories had been written about two of the departments at schools where Dr. Westfall trained, served, and taught: *The History of Pharmacology and Toxicology at West Virginia University* by Charles R. Craig and *Pharmacology at the University of Virginia School of Medicine* by Chalmers L. Gemmill and Mary Jeanne Jones. It was also encouraged by a quotation from the Hungarian-American biochemist Albert Szent-Györgyi, who won the Nobel Prize for Physiology or Medicine in 1937, the year that Dr. Westfall was born:

If you want to see ahead you must look back.

It is hoped that this book will be a valuable addition to the history of Saint Louis University and will serve as a permanent record of the accomplishments of the department and of the many men and women who have served and represented Saint Louis University over the years.

Information for this book, which became a labor of love for Dr. Westfall, came from multiple sources. There were numerous copies of correspondence from former chairs, deans, and faculty (including letters, memos, reports, and other writings) collected and assembled from the dean's office. Many of these were personally given to Dr. Westfall by Dr. William Stoneman Jr., Dean of the School of Medicine (1982–1994) and a well-known historian. Valuable information was also obtained from an unpublished manuscript on the history of the Department of Physiology at Saint Louis University prepared in 1966 by Dr. Florent E. Franke, Professor Emeritus and a faculty member from 1923 to 1963, which was passed on to Dr. Westfall by Dean Stoneman. Dr. Westfall also obtained some writings on the history of the school prepared by Dr. Alphonse M. Schwitalla, SJ, Dean of the School of Medicine (1927–1948). Archival material from the Medical Center Library and the Pius XII Memorial Library, including all of the bulletins and catalogs from the School of Medicine (1837–1855 and 1903–present), were examined, as well as excerpts from the Saint Louis University Libraries Digital Collections. Additional information was obtained from medical student yearbooks; university and Medical Center publications such as *Universitas*, *Front Page*, *St. Louis University Medical Center News*, *St. Louis University Medical Center MD's Line*, *Word from the School; of Medicine*, and *Parameters in Health Care*; countless obituary notices, historical encyclopedias, and educational materials (including course descriptions, objectives, and syllabi); curricula vitae of faculty; original articles from the medical literature; and information from *The Pharmacologist*, the quarterly news magazine of the American Society for Pharmacology and Experimental Therapeutics (ASPET), and *The Physiologist*, the newsletter of the American Physiological Society (APS), as well as personal recollections and experiences of the faculty and the author himself.

Acknowledgments

In addition to the people and sources mentioned in the preface, the following individuals, who have aided and assisted greatly in the success of this book, deserve special acknowledgment. First of all, special thanks to Kate Bax, Access Services Coordinator at the Medical Center Library, who provided all of the catalogs and bulletins from 1842 to 1855 and from 1903 to the present, as well as other archival material. These were indispensable in obtaining historical information. Special thanks also to Dr. Patrick McCarthy, Director of the Medical Center Library, and Professor John Waide, University Archivist Emeritus, for providing valuable information. Special thanks and acknowlegement is also given to Saint Louis University Archives and Digital Services for permission to use figures 1-10 and figure 12. As already mentioned, special thanks are due to Dean William Stoneman, Dr. Florent E. Franke, and the Reverend Alphonse M. Schwitalla for providing valuable information. The author would also like to thank Drs. David P. Westfall, Dean Emeritus at the University of Nevada, Reno, and Mark Voigt, the former interim Chair of the Department of Pharmacology and Physiology, Saint Louis University School of Medicine, for reading the original manuscript and providing valuable suggestions and editorship. Special thanks also to Linda Russell, who painstakingly typed the many versions of the manuscript. Nothing would have been possible without her patience and encouragement. I would also like to thank the many colleagues who have made suggestions. I offer my special thanks to Dr. Judith Siuciak, Executive Officer of ASPET, for providing photographs of Drs. John Auer and Erwin Nelson. Thanks also to Matthew Brown for help with photographs, Kenneth P. Parsley for the photograph of attendees, Thomas C. Westfall Symposium, Dr. Yoon Kim for the photograph of Dr. Adam's department, and Dr. Terrance M. Egan for the picture on the cover. I would like to thank Josh Stevens, publisher, and his colleague Barbara Northcott at Reedy Press for converting the book from manuscript to the end product. Finally, I would like to thank my family, and especially my wife, Gingy, for their support and encouragement over the years in what seemed at times an impossible task.

List of Tables

TABLE 1 Chairs of the Departments of Pharmacology and Physiology, Saint Louis University School of Medicine (Chapter 2)

TABLE 2 Key Dates Related to the School of Medicine (Chapter 2)

TABLE 3 Suggested Topics for Student Reports, October 1958 (Chapter 5)

TABLE 4 Extramural Support; Department of Physiology (1982–89) (Chapter 8)

TABLE 5 Representative Schedule, Medical Pharmacology Course (1988-89)

TABLE 6 Selected Topics in Neuropharmacology, late 1980s (Chapter 9)

TABLE 7 Selected Topics in Cardiovascular Pharmacology, early 1990s (Chapter 9)

TABLE 8 Selected Topics in Molecular/Biochemical Pharmacology, late 1980s (Chapter 9)

TABLE 9 Methodology Course, 1989 (Chapter 9)

TABLE 10 Annual Extramural Support, Department of Pharmacology (1979–90) (Chapter 9)

TABLE 11 Representative Funding of Faculty as PIs for 1987 (Chapter 9)

TABLE 12 Accomplishments of Faculty, Department of Pharmacology 1979–90 (Chapter 9)

TABLE 13 Organization and Administration, Department of Pharmacological and Physiological Science (as of 2003) (Chapter 10)

TABLE 14 Course Directors and Committee Chairs, Department of Pharmacological and Physiological Science (2001–02) (Chapter 10)

TABLE 15 Lecture, Review, and Exam Schedule of Medical Physiology Course PY705 (1992–93) (Chapter 10)

TABLE 16 Principles of Pharmacology Course Schedule (2012) (Chapter 10)

TABLE 17 Human Patient Simulator Worksheet (Chapter 10)

TABLE 18 Lecture Topics: Medical School Modules (Chapter 10)

TABLE 19 Responsible Conduct of Research, Phase 2: (Chapter 10)

TABLE 20 Representative Curriculum for the Graduate Program (Chapter 10)

TABLE 21 Advanced and Selected Topics in Pharmacological and Physiological Science: PPY511–514 (Chapter10)

TABLE 22 Syllabus for Human Physiology: PPY254 (Fall 2012) (Chapter 10)

TABLE 23 Physician's Assistants Program, Clinical Pharmacology (1991–92) (Chapter 10)

TABLE 24 General Physiology: PPYG504 (Fall 2001) (Chapter 10)

TABLE 25 Advanced Pharmacology in Primary Health Care Nursing: NRN508 (Chapter 10)

TABLE 26 Human Systems Physiology, Medical Anatomy & Physiology (Fall 2012) Chapter 10

TABLE 27 Representative Schedule and List of Topics for Drugs We Use and Abuse: BLA245 (Chapter 10)

TABLE 28 Extramural Support Generated by the Department of Pharmacological and Physiological Science (1991–2012) Chapter 10

TABLE 29	FY95 External Grants Report (Chapter 10)	**TABLE 40**	Medical Student Administration (2016–17) (Chapter 11)
TABLE 30	FY2000 External Grants Report (Chapter 10)	**TABLE 41**	Principles of Pharmacology Schedule, 2018 (Chapter 11)
TABLE 31	Nobel Laureates Who Gave Seminars in the Department (1979–2014) (Chapter 10)	**TABLE 42**	Course Schedule: Grant-Writing Course (Chapter 11)
TABLE 32	D'Agrosa Lectures (Chapter 10)	**TABLE 43**	PPY511, 512, and 513 (2015–16) (Chapter 11)
TABLE 33	Agenda of the Department of Pharmacological and Physiological Science Faculty Meeting, January 23, 2003 (Chapter 10)	**TABLE 44**	Extramural Support (2013–16) (Chapter 11)
TABLE 34	Department Retreats: Dates and Locations (Chapter 10)	**TABLE 45**	Department Pharmacology and Physiology Faculty Meeting Agenda, August 2, 2018 (Chapter 11)
TABLE 35	Summary of Accomplishments for the Department of Pharmacology (1979–90) and the Department of Pharmacological and Physiological Science (1990–2012) (Chapter 10)	**TABLE 46**	Department of Pharmacology and Physiology Retreat Agenda, May 9, 2019 (Chapter 11)
		TABLE 47	Pharmacology and Physiology Teaching Overview (2018–19) (Chapter 11)
TABLE 36	Former and Current Members of the Faculty Who Assumed Leadership Roles in Academia (Chapter 10)	**TABLE 48**	Departmental Service and Reputational Efforts (2018–19) (Chapter 11)
		TABLE 49	Departmental Research Overview (2018–19) (Chapter 11)
TABLE 37	Pharmacology and Physiology Departmental Faculty Meeting Agenda, January 27, 2017 (Chapter 11)	**TABLE 50**	Pharmacology and Physiology Strategic Plan (as of 2019) (Chapter 11)
TABLE 38	Winners of the Thomas C. Westfall Fellowship Award (Chapter 11)	**TABLE 51**	Total Number of PhD Graduates Mentored by Faculty in Pharmacology and Physiology at Saint Louis University (Chapter 11)
TABLE 39	Medical Student Teaching (2016–17) Year (Chapter 11)		

List of Figures

FIG. 1   Dr. William Beaumont
FIG. 2   Artist's rendition of the School of Medicine (ca. 1850) at Myrtle and Seventh Streets
FIG. 3   Dr. Daniel Brainard
FIG. 4   Dr. M. L. Linton
FIG. 5   Dr. Thomas Reyburn
FIG. 6   Dr. W. M. McPheeters
FIG. 7   Dr. Charles Alexander Pope
FIG. 8   Rev. W. B. Rogers, SJ
FIG. 9   Festus J. Wade
FIG. 10   Dr. Elias Potter Lyon
FIG. 11   Dr. John Auer
FIG. 12   Dr. Alphonse M. Schwitalla, SJ
FIG. 13   Dr. Irving Coret
FIG. 14   Dr. Erwin E. Nelson
FIG. 15   Dr. Elijah Adams and Department
FIG. 16   Dr. Budh Bhagat, Dr. René Wégria (chairman), and Dr. H. H. Oei
FIG. 17   Dr. Alrick Hertzman
FIG. 18   Article on Medical Horizons in the St. Louis Post-Dispatch, October 31, 1955
FIG. 19   Dr. Alexander Lind
FIG. 20   Department of Physiology (ca. 1981)
FIG. 21   Dr. Thomas C. Westfall
FIG. 22   Department of Pharmacology (ca. 1981)
FIG. 23   Extramural support per tenured faculty (1980–90)
FIG. 24   Department of Pharmacology, 1989
FIG. 25   Dr. Mary Ruh
FIG. 26   Letter from Dr. Stoneman announcing the formation of the Department of Pharmacological and Physiological Science in October 1990
FIG. 27   Advertisement of training program in the Department of Pharmacological and Physiological Science
FIG. 28   Extramural funding (1989–2012)
FIG. 29   Program for the 2012 Thomas C. Westfall Graduate Student Symposium
FIG. 30   Attendees: 2012 Thomas C. Westfall Graduate Student Symposium
FIG. 31   PhD graduates, 1981–90
FIG. 32   Graduate Students, 1990s
FIG. 33   Graduate students, ca. 1996
FIG. 34   Graduate students, 1999–2004
FIG. 35   Graduate students, 2007
FIG. 36   Graduate students, 2010
FIG. 37   Graduate students, 2011
FIG. 38   Graduate students, 2012
FIG. 39   Graduate students, 2015
FIG. 40   Graduate students, 2017
FIG. 41   Dr. Thomas Burris
FIG. 42   The Thomas Westfall Fellowship Award
FIG. 43   Dr. Mark Voigt
FIG. 44   Dr. Daniela Salvemini

CHAPTER ONE
THE FIRST SCHOOL of MEDICINE at SAINT LOUIS UNIVERSITY
1842–1855

Introduction and Background

The history of the departments of pharmacology and physiology as well as the history of the School of Medicine at Saint Louis University (SLU) can be divided into two phases: the early period, from 1842 to 1855, and the second period, from 1903 to the present time. From 1855 until 1903, Saint Louis University did not operate a medical school. The university itself came into existence in 1818, when the Most Reverend Louis Guillaume Valentin DuBourg founded the Saint Louis Academy. It became a college in 1820 and was chartered as a university in 1832. It is the oldest university west of the Mississippi River and the second oldest Jesuit University in the United States. [1, 2, 4–6]

The history of medical education at Saint Louis University actually began during the term of the university's first president, Peter Verhaegen, in 1835. Father Verhaegen asked the university's Jesuit trustees for permission to speak with local St. Louis medical doctors about forming a medical faculty for the university. Father Verhaegen was given permission and began discussion with Dr. Bernard G. Farrar, Dr. Hardage Lane, and Dr. Benjamin Boyer Brown, who were leaders in the recently organized St. Louis Medical Society, among others. Things moved quickly, and a board of trustees for the Medical Department was established by the university. This board included prominent residents from various professions, including Dr. Farrar; the Reverend William Greenleaf Eliot, who would found Washington University; John O'Fallon, a wealthy businessman and philanthropist; explorer and fur trader William Henry Ashley; entrepreneur John W. Johnson; H. L. Hoffman; Thomas P. Green; William Renshaw; and Judge M. P. Leduc. [2–5]

It was soon decided that the St. Louis Medical Society would help set up a medical school in cooperation with the university. An agreement was formalized in October 1836 between the university and the St. Louis Medical Society that the university would have a medical faculty.

Saint Louis University published a prospectus for its medical program in the university's annual catalog and bulletin for the 1836–37 academic year. The first medical faculty was selected by the St. Louis Medical Society of Missouri and approved by Saint Louis University in 1837. It consisted of the following: Dr. E. H. McCabe, Professor of Materia Medica (later called pharmacology); Dr. C. J. Carpenter, Professor of Anatomy and Physiology; Dr. William M. Beaumont, Professor of Surgery; Dr. Joseph Johnson, Professor of Theory and Practice of Medicine and Medical Jurisprudence; Dr. Hardage Lane, Professor of Obstetrics and Diseases of Women; and Dr. Henry King, Professor of Chemistry. [1, 5, 7] This school never got off the ground, and the 1840 Saint Louis University catalog stated, "The School was not operational." The school did give a diploma, however, to Benjamin Boyer Brown on August 7, 1839. Although it was only an honorary degree, it nevertheless was the first medical degree presented west of the Mississippi River. [5, 7]

The first six faculty were all distinguished individuals, but one in particular stands out above the others because of his contributions to science and physiology: Dr. William M. Beaumont. In 1990 the Reverend

Lawrence Biondi, President of Saint Louis University, and William Stoneman, MD, Dean of the School of Medicine, established the William Beaumont, MD, Professorship and Chair to be given to the Chair of Pharmacology and Physiology at Saint Louis University to honor Dr. Beaumont, a great teacher and leader in his field. Beaumont was the first named professor and chair of surgery at Saint Louis University, and he was known as the "Father of Gastric Physiology." The author of this book, Dr. Thomas C. Westfall, was the first William Beaumont, MD, Professor and Chair and holds the title of William Beaumont, MD, Professor and Chair Emeritus. For these reasons, a word or two about Dr. Beaumont seems warranted.

William Beaumont, MD (1785–1853)

The Beaumonts came from England to the American colonies in 1635, and William Beaumont's father, Samuel, and his uncles all fought in the Revolutionary War. Dr. William Beaumont, the second of Samuel and Lucretia Beaumont's nine children, was born on November 21, 1785, in Lebanon, Connecticut. William grew up on his father's farm and entered into an apprenticeship in medicine in the spring of 1811 under doctors Benjamin Chandler and Truman Powell in St. Albans, Vermont. [9, 10, 14, 15] This was common practice at the time, since there were only two medical schools in the United States: one in Philadelphia, which became the University of Pennsylvania, and one in New York, which became Columbia University. In June 1812 the Third Medical Society of Vermont approved William Beaumont to practice "Physic and Surgery." [9, 14, 15] He soon joined the Army of the North as a surgeon's mate with the 57th Regiment Infantry and served in Upper New York state in the border skirmishes of the War of 1812.[9] Following a period of private practice in Plattsburg, New York, he rejoined the army in 1819. On March 18, 1820, he was commissioned as Post Surgeon of the Army by President Monroe to Fort Mackinac, Michigan, in the northwestern frontier. His life was soon to change. In the late spring of 1822, a group of French-Canadian trappers arrived at Mackinac Island and were relaxing after hard winter months in the backwoods. Among them was a twenty-eight-year-old French-Canadian named Alexis Bidagan dit St. Martin. On the morning of June 6, St. Martin was wounded by the accidental discharge of a loaded gun while lounging in the local trading post of the American Fur Company. He was shot in the upper left abdomen. [9, 14, 15] Beaumont, the only physician for hundreds of miles, was called. According to Beaumont, the musket wound was "more than the size of the palm of a man's hand." It affected part of the lung, two ribs and the stomach. Dr. Beaumont treated the wound but remarked to bystanders, "the man cannot live for thirty-six hours." [9, 14, 15] To Beaumont and everyone else's astonishment, St. Martin continued to recover. Dr. Beaumont was repeatedly unsuccessful in fully closing the hole in St. Martin's stomach, and the wound became a permanent open gastric fistula, large enough that Beaumont could insert his entire forefinger into the stomach cavity. In 1825 Beaumont realized he'd been handed a remarkable opportunity because, at this time, almost nothing was known about the fate of ingested food. Since St. Martin was unable to work as a fur trapper, Beaumont made a deal with St. Martin that if he (St. Martin) would allow Beaumont to perform some observations and experiments on him he would hire him as his live-in handyman. St. Martin agreed, and Beaumont began his series of observations and experiments. Beaumont peered into St. Martin to see what the stomach was doing after different meals and collected samples of gastric juices for chemical analysis. The doctor became the first person to observe human digestion as it occurs in the stomach. By the time he was finished, Beaumont had carried out 178 experiments on St. Martin and had spent over five years studying the process of digestion. [9–15] In 1833 William Beaumont published his masterpiece,

FIG. 1 Dr. William Beaumont

entitled *Experiments and Observations on the Gastric Juice and the Physiology of Digestion.* [16, 17] This book was received as a clear, accurate, scientific work. [12, 15, 16, 18]

In July 1834 William Beaumont began service at his last army post, Jefferson Barracks, some three miles from St. Louis. [9] He, his wife Deborah Platt, and their children lived comfortably at the arsenal in St. Louis. They later moved to the city and he obtained permission to engage in private practice. He developed an extensive and lucrative practice. For a time, the Beaumonts shared quarters with Lt. Robert E. Lee (of Civil War fame) and his family. Lt. Lee was on assignment to supervise the dredging of the Mississippi River to improve navigation.

William Beaumont was offered the position of professor of surgery at the new Saint Louis University School of Medicine in 1837, but was unable to accept because of military obligations. He was refused a leave of absence and ordered to transfer to a post in Florida. [10] Rather than accept the transfer, he retired from the army on September 18, 1839. He continued the practice of medicine in St. Louis until his death in 1853. He was active in the St. Louis Medical Society and was elected its president on May 8, 1840.

In March 1853, while returning home from a visit with a patient late in the evening, he slipped on ice-covered stone steps, striking his head very forcefully. In a daze, he wandered about until met by a friend who accompanied him to his home. It was discovered that when he fell, he developed an occipital hematoma that later became infected. William Beaumont died on April 25, 1853, and was buried in Bellefontaine Cemetery in St. Louis. [9, 10]

Beaumont's legacy is large, and he has been called the "Father of Gastric Physiology." Garrison's *Introduction to the History of Medicine* states that Beaumont "was a true leader and pioneer in experimental physiology in our country." [7, 11, 18] In St. Louis, Beaumont High School, the Beaumont Hospital College of Medicine, the Beaumont Conference Room in Schwitalla Hall, and the William Beaumont, MD, Professor and Chair in Pharmacology and Physiology at Saint Louis University School of Medicine were all named after him.

Medical Education Continues in St. Louis

After the first attempt by Saint Louis University did not materialize, others tried to establish a medical school in the St. Louis area. Kemper Medical College (also known as McDowell's Medical College) was founded in St. Louis in 1839 by a young Kentucky-born doctor and surgeon named Joseph Nash McDowell. The college was located at the corner of Ninth and Cerre Streets, only a few blocks from Saint Louis University. This school gave the first course of medical lectures west of the Mississippi River and graduated its first doctor in 1840. It failed in 1845, and the medical department consequently transferred to the University of Missouri, where it remained affiliated for the next ten years. In 1855 it received an independent charter and officially became the Missouri Medical College. [5, 7, 8]

In 1841 steps were taken to establish a medical school at Saint Louis University on a firmer basis. Formal letters of acceptance of the professorships were read and approved. The board of trustees that would govern the medical department was appointed by the board and faculty of Saint Louis University, but, in order to free the department of all prejudice of a sectarian nature, it was agreed that the trustees would be selected from various religious persuasions as follows: two Unitarians, two Presbyterians, two Catholics, two Episcopalians, two Methodists, one Baptist, and one Reformed Presbyterian. In the event of the death, resignation, or removal of a trustee, his place would be filled by a person of the same denomination. [1, 2, 5]

The first lecture in the Medical Department of Saint Louis University was given to the students and a large public audience on March 28, 1842, by Professor Joseph W. Hall, MD. [5] The faculty during the first year of operation consisted of Dr. Joseph W. Hall, Dr. Daniel Brainard, Dr. Augustus H. Prout, Dr. Moses L. Linton, Dr. Joseph G. Norwood, Dr. Alvin Litton, and Dr. James V. Prather, who was not only head of surgery but also dean. [5]

In the fall, twenty-nine students from six states and one foreign country began their instruction in a small house owned by Dean Prather on the north side of Washington Avenue, adjacent to the university. There

FIG. 2 Artist's rendition of the School of Medicine (ca. 1850) at Myrtle and Seventh Streets

were at this time only eight other universities in the United States with functioning medical departments (Pennsylvania, Columbia, Harvard, Maryland, Yale, Willoughby, Transylvania, and New York) and just one chartered college, Dartmouth. The *Boston Medical and Surgical Journal* welcomed SLU to its fold with this notice in 1842:

> *St. Louis University–The Medical Department appears to be well sustained and the school is flourishing as any on the Atlantic shore. Dr. Linton, Professor of Obstetrics, gave an introductory lecture at the opening of the regular annual term in the early part of November, which was well received and the doctor alluded to the advantages of St. Louis for both public and private teaching and concluded that no city west of the mountains affords greater facilities and more ample means for medical instruction.*

The first six medical degrees were awarded in 1843.

Departments of Physiology and Materia Medica (Pharmacology)

The history of the departments of pharmacology and physiology can be divided into the same two phases as that of the medical school: 1842 to 1855 and 1903 to the present. From 1842 to 1855, there were four chairs of the Department of Physiology: Dr. Daniel Brainard (1842–44), Dr. Joseph W. Hall (1844–46), Dr. Henry Bullitt (1846–48), and Dr. R. S. Holmes (1848–55).[1] There were also four chairs of the Department of Pharmacology, or Materia Medica, as it was called at the time: Dr. H. Augustus Prout (1841–42), Dr. J. G. Norwood (1842–46), Dr. Thomas Reyburn (1847–1850), and Dr. W. M. McPheeters (1850–55).[1]

Early Chairs: Department of Physiology

Daniel Brainard, MD (1842–44)

Dr. Daniel Brainard was the first professor of anatomy and physiology at the Saint Louis University School of Medicine. Dr. Brainard graduated from Jefferson Medical School in 1832 and studied languages and sciences. From 1834 to 1839 he was a faculty member at a medical school in Chicago. He studied in Paris from 1839 to 1841 and occupied the chair of physiology at Saint Louis University from 1842 to 1844. He left for Rush Medical College in Chicago in 1844 and is considered the founder of that school. He was appointed chair of anatomy and surgery and served in that position until his death in 1866. He helped found the *Illinois Medical and Surgical Journal* (which would ultimately become the *Chicago Medical Journal*) in April 1844. He helped establish the first general hospital in Chicago in 1847. Among his scientific accomplishments was a bone drill that was in clinical use until 1936. He read papers before the French Academy of Sciences in Paris and the Société de Chirurgie, of which he was a corresponding member. Brainard received a prize from the American Medical Association for his essay on fractures (ca. 1853). Dr. Brainard died of cholera on October 10, 1866, at the age of fifty-four, one day after giving a lecture on cholera. [7]

Joseph W. Hall, MD (1844–46)

Dr. Hall was professor of theory and practice of medicine and clinical practice at Saint Louis University School of Medicine (1842–44) and succeeded Dr. Brainard as chair of physiology in 1844. He served in this capacity for two years as professor of physiology, pathology, and clinical practice. He published an article on consumption in the *St. Louis Medical and Surgical Journal*, considered the first medical journal to be published west of the Mississippi River. The journal was founded in 1843 by Dr. M. L. Linton, a native of Kentucky who was educated professionally in the East, in Paris, and in Edinburgh, and was the first professor of obstetrics in the Medical Department of Saint Louis University. In 1845 the journal was enlarged, and Dr. William McPheeters—who later became chair of pharmacology (1850–55)[1, 19]—and Dr. Fourgeand, who were both on the medical faculty, joined Dr. Linton as editors. [19] Several other members of the medical faculty were said to have been involved in editing and contributing to this journal. Dr. Hall was a much sought-after lecturer at both Kemper (McDowell's) Medical College and Saint Louis University. [1]

Henry M. Bullitt, MD (1846–48)

Dr. Bullitt succeeded Dr. Hall and was elected to the chair of physiology and pathology in 1846 and served until 1848. He delivered two courses of lectures during this period and published two articles on medical education in the *St. Louis Medical and Surgical Journal*. He was appointed as a delegate to the National Medical Conventions on behalf of Saint Louis University and was also asked to represent the medical department of the state university. Following his stay at Saint Louis University, he gave introductory lectures at Transylvania University in 1849 and 1850, as well as at the Kentucky School of Medicine in 1853 and 1854 and again in 1865. He also gave the introductory lecture at the opening session of the Medical Department of the University of Louisville in 1867 and was editor of the *Kentucky Medical Recorder* and the *Transylvania Medical Journal*.

The *Annual Catalogue and Announcement of the Medical Department of the St. Louis University* stated: "The high reputation as a scholar and medical writer employed by the lately selected Professor of Physiology and Pathology justifies the anticipation of a complete course on these highly interesting branches, in which all facts appertaining to them are known and will be taught." [1]

R. S. Holmes, MD (1848–55)

Dr. Holmes graduated from Jefferson Medical College in 1838 and then visited various medical centers in Europe. He scored high in an examination to become an assistant surgeon in the US Army and was appointed to the position. He was appointed professor of physiology and medical jurisprudence at Saint Louis University School of Medicine from 1848 until his death in 1855. Dr. Holmes was a frequent contributor to *The American Journal of the Medical Sciences* and the

FIG. 3 Dr. Daniel Brainard

FIG. 4 Dr. M. L. Linton

St. Louis Medical and Surgical Journal, as well as many literary journals. The papers of Dr. Holmes on climate, malaria, and quinine published in the *St. Louis Medical and Surgical Journal* were widely quoted in the East, and some were reproduced at length in foreign publications. He became paralyzed in August 1854 and died June 26, 1855.

J. H. Watters, MD, Acting Chair (1854–55)

After Dr. Holmes became paralyzed in 1854, Dr. J. H. Watters took his place as acting chair until Saint Louis University discontinued its operation of the medical school in 1855. Dr. Watters wrote an essay entitled *The Doctrine of Life*, which was said to have caused much controversy.

SYNOPSIS OF PHYSIOLOGY COURSE FROM LECTURES OF PROFESSOR HOLMES

The aim of the Professor in this department is to present to his class the latest views in the science, and chiefly in this manner in which these views have been modified from year to year by subsequent investigators.

He divides the subjects of enquiry into three great divisions. The first comprehends Comparative or Modified Life. Under this head the terms Life, Physiology, Biology, Zoology, Comparative Anatomy, are defined and classified in their relative bearings. Vegetable Physiology is dwelt upon in its connection with life-differences between organic and inorganic matter-elementary parts of animals, plants and inorganic matter-man, his position in the scale of animated being - Ethnology-cessation of this life, called death.

The second division includes Animal or Relative Life, generally placed in its duties or functions under the dominion of the will, opposed to vegetable life. This form of life, in all its phases, chiefly considered in reference to man. Sensibility, including the physiology of the nervous system, muscular motion, language or speech, reproduction or generation; reasons for placing the latter in this division.

Thirdly: Nutritive or Organic Life, in its duties or functions, placed beyond the domination of the will, and frequently analogous to vegetable life. Considered chiefly in reference to man. Circulation, blood, respiration, exosomes and endosomes more thoroughly discussed; digestion, with resume of experiments thereon; absorption; practical acquaintance with the microscope; cell theory more completely considered; nutrition, secretion, calorification.

The course is well illustrated by magnified drawings, preparations, sketches upon the black board, experiments upon living animals, etc.

Among other means of further illustration of the course is a powerful microscope, by Ross of London, capable of magnifying 1300 diameters. This instrument forms an invaluable aid in elucidating minute form and structure, and in order that every student may have a full opportunity of inspecting its wonders and of becoming practically familiar with its use, the Professor is in the habit of making subdivisions of the class, and of meeting them at extra hours in his room for this especial purpose.

Students will also have the advantage of the experiments of Professor Blake on the physiological action of medicines, on which branch of Physiology he has been engaged for several years, at the request of the British Association for the Advancement of Science. The more important phenomena connected with the circulation of the blood will be fully illustrated by the use of the Hemadynamometer.

At the conclusion of the course on Physiology, the subject of medical jurisprudence will receive that attention which its importance demands.

Text Books: Carpenter, Kirke, or Muller.

Synopsis of the Course in Physiology as Taught by Professor Holmes

A synopsis of the physiology course as summarized in the *Annual Catalogue and Announcement of the Medical Department of the Saint Louis University* for the 1849–50 session appears on page 6. [1] It is representative of all the courses taught from 1842 to 1855.

On several occasions, Professor Holmes traveled to Europe. On one of these trips, it was stated on several occasions in the *Annual Catalog* of the time,

> *Professor Holmes will go on business connected with the school to Europe and will fully equip himself in everything pertaining to the Chair of Physiology in the way of books, plates, models and preparations, besides obtaining numerous other articles for the gen museum. Amongst these will be a powerful microscope in order to more completely elucidate minute form and structure.* [1]

Early Chairs: Department of Pharmacology (Materia Medica)

H. Augustus Prout, MD (1841–42)

The first professor of materia medica (later called pharmacology) was H. Augustus Prout, MD. He actually served as professor of chemistry, materica medica, and botany for only one year. [1]

J. G. Norwood, MD (1842–46)

Dr. Norwood succeeded Dr. Prout and was professor of materia medica, therapeutics, and jurisprudence. He served from 1842 to 1846. The *Annual Catalogue* stated that "it is entirely needless to say that his course is not surpassed either in content or in extent of learning, by any other in the same Department in the Country." [1]

Thomas Reyburn, MD (1847–50)

Dr. Reyburn succeeded Professor Norwood and served as the third professor of materia medica and therapeutics from 1847 to 1850. He published several articles in the *St. Louis Medical and Surgical Journal*, including one entitled "On the Use of Thepine on Some Varities of Fracture of the Cranium." A synopsis of his course as summarized in the *Annual Catalogue* for the 1849–50 session follows on page 8. [1] It is representative of all the courses taught from 1842 to 1855.

W. M. McPheeters, MD (1850–55)

The last professor of materia medica and therapeutics prior to the closing of the first School of Medicine at Saint Louis University was Dr. W. M. McPheeters. He served as director of pharmacology from 1850 to 1855. Dr. McPheeters published a paper in the *St. Louis Medical and Surgical Journal* entitled "History of the Cholera Epidemic of 1849" that attracted worldwide attention. He became an editor of the *St. Louis Medical and Surgical Journal* in 1845.

FIG. 5 Dr. Thomas Reyburn

FIG. 6 Dr. W. M. McPheeters

Synopsis of Pharmacology Course from Lectures of Professor Reyburn

The lectures of the branch embrace the description of remedial agents—the general Principles of Therapeutics—the indications which medicines are capable of fulfilling and the laws governing the application of remedies.

The different agents of the Materia Medica—their source, their physical, chemical and commercial history are noticed and the methods of distinguishing the superior from the inferior qualities, and adulteration are enumerated. The various official formulae are described and the combinations demanded in prescriptions are detailed in length.

The articles of the Materia Medica are classified according to the prominent therapeutic effects ordinarily induced and the special indications governing their prescriptions are discussed. The classes are subdivided into groups, the individual members of each group being examined in the order of their relative value and importance. The precise nature of their action is described and the special effects of the different members of each group are compared so as to bring into relief the circumstances which must in prescription guide the practitioner in his selection. The modification represented by disease idiosyncrasy, habit, epidemic constitutions etc on the effects of the remedies are fully noted, as well as the various diseases in which each remedy is recommended, the circumstances under which it is to be administered, the indications and contra-indications to its use, its form of exhibition, incompatibilities etc.

Medicinal agents in their relationship as poison are elaborately discussed and their effects, method of detection and treatment are detailed.

The influence of therapeutic agents on the function, tissues and fluids of the economy are fully noticed; in short, an effect is made to keep pace with the progressive improvements of medical philosophy and to reconcile, so far as may be, therapeutic reasoning with conclusions legitimately based on the more recent disease discoveries in physiology and pathology. Whilst on the one hand a visionary regard to speculative theory is condemned, it is equally the object of the lecturer to avoid the opposite error of attachment to fallacious experience and blind empiricism.

To demonstrate his course, the lecturer avails himself of an extensive collection of large and finely colored plates and dried specimens of the articles of Materia Medica. He will have an opportunity of exhibiting to the class many foreign and indigenous plants in their growing state.

Textbooks recommended to the student in this department are:

- Traité de Thérapeutique et de Matière Médicale by Trousseau and Pidoux
- The Dispensatory of the United States
- A Dispensatory, or Commentary on the Pharmacopoeias of Great Britain and the United States by Christison, edited by Griffith
- The Elements of Materia Medica and Therapeutics by Pereira
- General Therapeutics and Materia Medica by Dunglison
- Mayne's Dispensatory and Formulary by Griffith [1]

Separation of Saint Louis Medical College from Saint Louis University

During the late 1840s and early 1850s, there were a number of hostile demonstrations in an attempt to arouse a mob movement against the university and its medical school. This was largely due to the continued nativist, anti-Catholic, anti-foreign movement that was on the rise in St. Louis and other parts of the country. These groups were typically called "Know-Nothings," as that was their standard response to any questions about their activities. In reality this was the outgrowth of decades of festering resentment against foreigners. In St. Louis, much of the Know-Nothing movement, or nativist anger, was directed toward the growing number of Irish Catholic immigrants. Attempts by the faculty to sever relations between the medical school and the university started in 1848, but the break was not approved by the university trustees. However, in 1854 a majority of the medical faculty once again voted to separate from the university due to ongoing problems with the Know-Nothing movement. Father John Verdin (a native of St. Louis with French and Irish ancestry), the president of Saint Louis University, called together the university trustees, and this time they approved the separation. [3, 4, 5]

Although the situation was regrettable, the separation was by mutual consent and without unfriendly feeling on either side, especially since the peculiar circumstances of the times seemed to compel the medical department to adopt this course. Underscoring the fact that the parting was an amicable one, the university conferred honorary degrees on Dr. Pope and Dr. Linton at commencement ceremonies on June 18, 1855, when forty doctors of medicine were awarded. During the medical department's existence from 1842 to 1855, the university conferred 317 medical degrees. More than a few of these alumni went on to distinguished careers in medicine and other fields. [3, 4, 5]

During the last year of operation, there were 141 medical students at Saint Louis University School of Medicine. The remaining school was then known as St. Louis Medical College.

It was also known as Pope's College, after Dr. Charles Alexander Pope, who joined the faculty in 1842, became head of surgery in 1844, and was named dean of the Medical Department in 1847. Dr. Pope had studied under Daniel Drake in Cincinnati and Samuel Gross at the University of Pennsylvania, where he received his medical degree at the age of twenty-one. He then studied in London, Edinburgh, and Paris, where he and Linton were classmates in 1839. Charles Alexander Pope later was elected president of the American Medical Association. He was the first Westerner to head the Association and, at age thirty-six, the youngest person ever to hold that prestigious office. St. Louis Medical College existed under that name from 1855 until 1891, when it merged with Missouri Medical College (1845–91), itself known as Kemper Medical College or McDowell's Medical College before 1845. In 1891, the two schools became the Washington University Medical Department and later the Washington University in St. Louis School of Medicine. [2, 4, 5]

FIG. 7 Dr. Charles Alexander Pope

Chapter Two
FORMATION of the CURRENT SCHOOL of MEDICINE at SAINT LOUIS UNIVERSITY

1903–Present

The second phase of the history of Saint Louis University School of Medicine begins in 1903, when the university acquired by purchase the Marion Sims-Beaumont College of Medicine. The Beaumont Hospital Medical College, named in honor of the famous frontier physician and physiologist William Beaumont, was founded in 1886 at Sixth and Walnut Streets in St. Louis. It moved to the southwest corner of Jefferson and Pine in 1890. The Marion Sims College of Medicine was formed by Dr. Young H. Bond in 1890 and named for Dr. Marion Sims, who was known as the "Father of Gynecology." It was located twelve blocks south of Saint Louis University across Mill Creek Valley in the vicinity of Grand and Caroline. Marion Sims College opened Rebekah Hospital in 1893 for clinical and bedside instruction, and the following year established a dental department. On May 1, 1901, the Marion Sims College of Medicine was consolidated with Beaumont Hospital Medical College. Both of these schools were owned and operated by a group of St. Louis's finest physicians. The consolidation was the result of an effort on the part of the two schools to unite into one large institution for the purpose of strengthening the advantages each offered. Both had maintained a large faculty with equipment and facilities that were amply sufficient for the stage of medical education each represented. The consolidated school, however, would offer opportunities for medical instruction that each alone could never provide. The equipment of the building on Jefferson and Pine was transferred to the Grand-Caroline campus. [5,7]

During the presidency of Father William Banks Rogers, who was president from 1900–1908 and often referred to as the "second founder" of Saint Louis University, plans were initiated for the integration of a new medical school into the university. Father Rogers began looking for a medical school in 1902. Rogers sent Father James Brent Matthews to discuss the idea of purchasing the Marion Sims-Beaumont College of Medicine with Dean Young H. Bond. Festus J. Wade, an Irish-born civic leader and the president of the Mercantile Trust Company, reassured Father Rogers that another

FIG. 8 Rev. W. B. Rogers, SJ

FIG. 9 Festus J. Wade

medical school was needed in the metropolitan St. Louis region and helped secure the $115,000 needed for the purchase.

Father Rogers began tramping the streets of St. Louis collecting the money. According to Father W. B. Faherty, "this was a long, tiring tedious process." [5] Substantial benefactors were identified. Besides Festus Wade himself, others were Daniel Walker, John Scullin, Daniel G. Nugent, Richard Kerens, Jane Lindsey, Mary Scanlan, Mrs. E. Walsh, and George Backer. In addition, $5,000 came from the estate of a Jesuit scholastic, John L. McNichols. The fundraising drive netted half the amount, and the sale of the remaining sections of the Clark estate in north St. Louis provided the other half. [5]

The decision by Saint Louis University to purchase the Marion Sims-Beaumont School of Medicine was supported by the Council of Medical Education and Hospitals of the American Medical Association which recommended that all medical schools should have a university affiliation.

According to Father W. B. Faherty in his book *Better the Dream: Saint Louis University and Community, 1818–1968*, historian Gilbert Garraghan concluded many years later, "The re-establishment of the faculty of medicine was … a turning point in the University's history." For instance, in the graduation of 1905, sixteen men received the degree of bachelors of arts, while ninety-three received the degree of doctors of medicine. [5]

As described in the university's annual catalog of the time:

> *To develop a true university school of medicine it is essential that the fundamental departments of medicine be placed on the same plane as other university branches. This requires that anatomy, chemistry, physiology, pathology, bacteriology and pharmacology be taught by specialists who devote their time exclusively to teaching and research.*

As stated in the 1903–04 catalog, "This plan inaugurated by the university during the session of 1903–04 has now been accomplished and the instruction of the first two years is placed permanently upon a University basis." It was further stated in these early catalogs "that the University and its work have been rapidly improved in every possible way and this has resulted in a large and vigorous school adequately equipped for the best and highest kind of medical education." [6] The most notable advancements were:

> 1) *The requirement for admission to the courses has been steadily raised.*
>
> 2) *The library, laboratories and other equipment have been greatly increased.*
>
> 3) *The courses of study have been strengthened.*
>
> 4) *In the fundamental branches, anatomy, chemistry, physiology, pathology, bacteriology and pharmacology, the corps of lecturers has been augmented by eminent specialists who take charge of these departments and who are exclusively devoted to their work.*
>
> 5) *The clinical facilities have been doubled and now offer the students extraordinary advantages, probably unequalled in the country.*

After the current School of Medicine was formed in 1903, there was a single Department of Pharmacology and Physiology from 1903 to 1921. There were three chairs, or directors: Dr. Charles Shattinger (1903–04), Dr. Elias Potter Lyon (1903–13), and Dr. Don R. Joseph (1913–21). From 1921 until 1990, the Departments of Pharmacology and Physiology were separate. In October 1990 the departments were merged once again under the name of Pharmacological and Physiological Science and were later renamed Pharmacology and Physiology. The chairs of Pharmacology were: Dr. John Auer (1921–1948), Dr. Robert H. K. Foster (1948–52), Dr. Erwin E. Nelson (1952–57), Dr. Elijah Adams (1958–63), Dr. René Wégria (1963–79), and Dr. Thomas C. Westfall (1979–90). Dr. Irving Coret served as interim chair from 1957 to 1958.

The chairs of physiology were Dr. Don R. Joseph (1921–28), Dr. Alrick B. Hertzman (1928–1966), and Dr. Alexander Lind (1968–88). Dr. Leo Senay acted as interim chair from 1966 to 1968, and Dr. Mary Ruh served as interim chair from 1988 to 1990.

As mentioned, the Departments of Pharmacology and Physiology merged in October 1990. Dr. Thomas C. Westfall was appointed chair and was named the first William Beaumont Professor and Chair of the newly formed Department of

TABLE 1 Chairs of the Departments of Pharmacology and Physiology, Saint Louis University School of Medicine

Original School (1842–1855)

Pharmacology
- Dr. H. Augustus Prout (1841–42)
- Dr. J. G. Norwood (1842–46)
- Dr. Thomas Reyburn (1847–50)
- Dr. W. M. McPheeters (1850–55)
- Dr. J. H. Watters (Acting Chair) (1855)

Physiology
- Dr. Daniel Brainard (1842–44)
- Dr. Joseph Hall (1844–46)
- Dr. Henry Bullitt (1846–48)
- Dr. R. S. Holmes (1848–55)

Current School (1903–present)

Pharmacology and Physiology
- Dr. Charles Shattinger (1903–04)
- Dr. Elias Potter Lyon (1904–13)
- Dr. Don R. Joseph (1913–21)

Pharmacology
- Dr. John Auer (1921–48)
- Dr. Robert Foster (1948–52)
- Dr. Erwin Nelson (1952–57)
- Dr. Irving Coret, Interim Chair (1957–58)
- Dr. Elijah Adams (1958–63)
- Dr. René Wégria (1963–79)
- Dr. Thomas Westfall (1979–90)

Physiology
- Dr. Don Joseph (1921–28)
- Dr. Alrick Hertzman (1928–66)
- Dr. Leo Senay, Acting Chair (1966–68)
- Dr. Alexander Lind (1968–88)
- Dr. Mary Ruh, Acting Chair (1988–90)

Pharmacology and Physiology
- Dr. Thomas C. Westfall (1990–2013)
- Dr. Thomas Burris (2013–18)
- Dr. Mark Voigt, Interim Chair (2018–19)
- Dr. Daniela Salvemini, Interim Chair (2019–present)

TABLE 2 Key Dates Related to the School of Medicine

1818	St. Louis Academy was founded, becoming the first institution of higher learning west of the Mississippi River
1820	Name changed to St. Louis College
1832	Name changed to Saint Louis University and chartered by the state of Missouri, becoming the first university west of the Mississippi River
1837	Early medical school proposed and first six professors named
1839	The school awards its first diploma, the first medical degree given west of the Mississippi River
1840	Kemper (McDowell's) College of Medicine opens
1842	Early Saint Louis University School of Medicine officially becomes operational
1845	Kemper (McDowell's) College of Medicine closes and becomes University of Missouri College of Medicine
1855	University of Missouri College of Medicine becomes the Missouri College of Medicine
1855	Early Saint Louis University School of Medicine closes and becomes St. Louis College of Medicine
1886	Beaumont Hospital Medical College formed
1890	Marion Sims Medical College formed
1891	Missouri Medical College and St. Louis Medical College become Washington University School of Medicine
1901	Beaumont Hospital Medical College and Marion Sims College of Medicine merge
1903	Marion Sims-Beaumont College of Medicine becomes the current Saint Louis University School of Medicine

Pharmacological and Physiological Science. He served from 1990 to 2013. Dr. Mark Voigt served as interim chair from January 2013 to July 2013. Dr. Thomas Burris was then appointed the second William Beaumont Professor and Chair and served until February 2018. Dr. Mark Voigt was again appointed interim chair until his retirement in 2019. At this time, Dr. Daniela Salvemini was named interim chair. In 2014 the name of the department was changed to Pharmacology and Physiology.

Table 1 summarizes the chairs of the Departments of Pharmacology and Physiology, Saint Louis University School of Medicine in the original school (1842–1855) and in the current school (1903–present), and Table 2 shows dates of key events relevant to the School of Medicine.

Chapter Three
DEPARTMENT of PHARMACOLOGY & PHYSIOLOGY: The SHATTINGER, LYON, & JOSEPH ERAS

1903–1921

The Shattinger Era (1903–04)
Biographical Sketch of D. Shattinger, MD

After Saint Louis University purchased the Marion Sims-Beaumont College of Medicine in 1903, Dr. D. C. Shattinger became the first chair of the Department of Pharmacology and Physiology at the new Saint Louis University School of Medicine. Dr. Shattinger was born in 1865 and received his doctor of medicine from the St. Louis Medical College (which was the Medical Department of Washington University in 1903). Dr. Shattinger was professor of Physiology at the Beaumont Hospital Medical College from 1892 to 1903. He became professor of Pharmacology and Physiology at Saint Louis University Medical College from 1903 to 1904. Dr. Shattinger wrote several papers, chiefly on clinical subjects, although one was entitled "Practical Physiology of the Stomach." He later did postgraduate work at Harvard University and became an eye, ear, nose, and throat specialist. He was professor and chair for only one year. [6, 7]

Members of the Department
- Charles Shattinger, MD, Professor
- L. E. Burford, MD, Assistant

Course in Physiology
- Freshman Year–First Semester
  Lectures: Professor Shattinger
  2 hours/week
  Exercises: Professor Burford
  1 hour/week
  Laboratories: Professor Shattinger
  1 hour/week
- Sophomore Year–First Semester
  Lectures: Professor Shattinger
  2 hours/week
  Exercises: Professor Burford
  1 hour/week
  Laboratories: Professor Shattinger
  1 hour/week

The Lyon Era (1904–13)
Biographical Sketch of Elias Potter Lyon, PhD

Dr. Lyon was the second chair of the Department of Physiology as well as the second chair of Pharmacology at Saint Louis University School of Medicine. The department's name was actually Physiology, Physiological Chemistry, and Pharmacology. He served as chair of Physiology for nine years, from 1904 to 1913, and Pharmacology for ten years, from 1904 to 1914. He also served as dean of the School of Medicine from 1907 to 1913. Dr. Lyon received his bachelor of arts and bachelor of science at Hillsdale College in 1891 and 1892, respectively, and his doctor of philosophy from the University of Chicago in 1897. He was an instructor at Bradley Polytechnic Institute for a few years and at the Marine Biological Laboratory in the summer for five years. He was assistant professor of physiology at Rush Medical College from 1900 to 1904 and assistant professor of physiology and assistant dean at the University of Chicago from 1901 to

FIG. 10 Dr. Elias Potter Lyon

1904. He served as professor of physiology at the Saint Louis University School of Medicine and was director of the department from 1904 to 1913. He also had responsibility for the Department of Pharmacology. This was followed by the position of dean and professor of physiology at the University of Minnesota School of Medicine from 1913 to 1936. Dr. Lyon wrote many papers on experimental subjects early in his career and later became very involved in medical education. He was a leader among the deans in the American Association of Medical Colleges (AAMC) and was elected president of the association in 1913 and 1914. He is credited with converting the University of Minnesota from an old-style school to a modern one in both teaching and research. Dr. Lyon died in 1937.

Aims of the Department during the Lyon Era

The aims of the department as stated in the *Bulletin of Saint Louis University: Announcements of the School of Medicine* were:

> *The Department of Physiology will attempt to ground the students thoroughly in the classical knowledge of the Chemistry and Physics of the human body, recognizing that the mass of facts is essential to medicine. The knowledge will be taught largely by the laboratory method. The students will learn to do experiments for himself, thus receiving only firsthand information, but also training for the future work, whether investigational or purely technical. The aim is to present the practical side; but not merely as a memory drill. Rather the practical as founded on and the outcome of scientific principles is the ideal which is kept in view in the teaching of this subject.*
>
> *It is recognized, however, that little of our traditional "Human Physiology" is really founded on the study of the human subject. Most of it is derived from experiments on animals. Much has been learned from the higher forms, such as mammals, but very often phenomenon is most clearly exemplified by the very lowest organisms. Therefore, the subjects of comparative and general physiology will be emphasized. The chemistry of living matter, together with its foods and wastes, will be taught both from the scientific and the technical standpoint. The student will learn not only the convenient medical tests, but also the body processes which the tests exemplify.*
>
> *The direction of current research will be pointed out, and the student so guided that he may follow the progress of physiology and pharmacology and understand the application to medical science which seems sure to follow in the next few years. Research will be fostered and all who have the training and aptitude for original work will be given the opportunity and encouragement.*
>
> *The laboratories for Physiology, Physiological Chemistry and Pharmacology occupy nearly two floors of the new building. They have recently been fitted out with all apparatus needed for a thorough modern course for medical students. Small laboratories for research work have also been established. Trained investigators are in charge of both teaching and the research laboratories. A library has been started and most of the physiological journals are subscribed for.* [6]

Synopsis of the Course

Physiology and Physiological Chemistry

1. Physiology of Blood, Circulation, Respiration, Digestion, and Absorption. Second semester freshmen Year. Lectures 2 hrs/week; recitation 1 hr/week. (Professor Lyon). Laboratory work; 4 hrs/week (Professor Lyon, Mr. Brown, and Assistants).
2. Physiology of Excretion, Metabolism, Muscle, the Nervous System, and Senses. First semester, sophomore year. Lectures 2 hrs/week; recitation 1 hr/week (Professor Lyon).
3. Chemistry of the Cell, including Enzyme Action. Second semester, freshman year. Lectures 2 hrs/week; recitation 1 hr/week. Lab work 4 hrs/week (Professor Neilson).

4. Chemistry of Excretions; chiefly urine. First semester, sophomore year. Lectures 2 hrs/week; recitation 1 hr/week. Lab work 2 hrs/week (Professor Neilson).
5. Conferences. 1 hr/week during second half of freshman year and first half sophomore year. These will be devoted to the presentation and discussion of papers by students. Each student will be assigned a subject with references to the original literature; which he must consult and critically review.

Courses in Pharmacology
6. Pharmacology. First semester, junior year Lectures, demonstrations and recitations 3 hrs/week (Mr. Brown); Laboratories; 6 hrs/week.

OVERVIEW OF PHYSIOLOGY
Physiology begins in the second semester of the freshman year and therefore after the students have had their Histology and Comparative Anatomy and some part of Chemistry. The students are first introduced to the Physiology of Muscle, Circulation, Blood and Respiration.

In the first semester of the sophomore year, the Physiology of the nervous system, for which the student has been prepared by his cat anatomy and by his Neurology, is taken up followed by Physiology of the senses. About the middle of the first semester of the sophomore year, the study of secretion, digestion, absorption, excretion and metabolism is begun and carried forward coordinated in Physiology and Physiological Chemistry, the former considering more the physical and nervous aspects of the phenomena; the latter the purely chemical side.

PHYSIOLOGY PROPER
As mentioned, Physiology is begun in the second semester of the freshmen year. Two lectures, one recitation, and six hours of laboratory work are given each week. Beginning with the subject of muscle and contractibility in general, the student is led to see that all external manifestations of life take the form of motion. In the laboratory, it is not believed to be profitable for the medical student to go into the graphic study of muscle to any great extent. Enough experiments are given to acquaint him with the more important phenomena and make him reasonably familiar with apparatus of methods.

The circulation of the blood is studied largely on mammals. The students work six at a table and so each one has an active part in each experiment, one serving as anaesthetist, a second preparing the electrical apparatus, a third acting as operator, etc. The students open the chest and study the living heart; they make blood pressure experiments; they study the effect of stimulation of various nerves on the heart and blood vessels. Great use is made of Porter's artificial scheme for making clear the physics of the circulation. The human pulse and heart beats are graphically recorded. The blood is studied chiefly from the physical and functional point of view. Each student learns how to make the various tests for blood; quantitative and qualitative tests for haemoglobin, including the use of various haemoglobinometers and the spectroscope.

Under respiration are considered chiefly the physical and nervous phenomenon of external respiration, the actual use of oxygen in the body being left for consideration under metabolism. The usual graphic experiments are thrown into the background for some introductory work in percussion and auscultation and for experiments on the effects of carbon dioxide, heat and cold on the respiratory center. A final examination on the work of this semester brings the Freshman Physiology to a close.

In the first semester of the Sophomore year, Physiology is continued beginning with the nervous system. The laboratory work on nerve, which is in most schools considered in connection with muscle, is here developed both in lecture and laboratory in connection with the central nervous system. It is believed

the standpoint of the anatomical and functional relationships of the peripheral nerves to the central system, but also for the practical reasons that the nerve experiments if given in the first year would crowd out more important work on the circulation, blood and respiration. Certain experiments, such as determination of reaction time, the behavior of the reflex frog, the stimulation of the cerebral cortex and of the anterior and posterior roots of the spinal nerves in mammals and the removal of parts of the brain complete the laboratory work on the nervous system.

About five weeks are devoted to the special senses. The lectures take up both theoretical and practical phases of the subject. In the laboratory Kuehne's artificial eye is used to make the student better acquainted with the problems of refraction met with in vision. The use of the ophthalmoscope (at least on the Thorington eye), perimetry, and color blindness are other practical topics taken up in the laboratory. In the Physiology of digestion and secretion effort is made to keep pace with the great work now being done in Pawlow's laboratory in St. Petersburg and by certain English investigators.

A part of the Physiology course is the reading by students of important articles concerning new investigations. Each student receives a subject and references. He must look up the latter in the library and write up a report. The best reports presented are read before the class. It is felt that this work serves a variety of useful purposes. It helps the student to see that text books are only compilations and that the sources of these are the various archives and journals where new discoveries are reported. It gives him some acquaintance with physiological investigations at firsthand and shows him how such work is carried on and made known to the scientific world. It tends to develop the critical faculty by compelling him to compare views and work of various investigators upon given subjects. It gives him some notion of the evolution of the science, as he studies the development of the knowledge of a given topic through a series of papers. Finally it affords him some valuable experience in writing up a subject from an investigation of the literature.

The examinations in Physiology consist of: a) a laboratory examination in which the student is called on to repeat unaided an experiment in the course; b) a written examination; c) an individual oral examination. The final grade is founded on the results of these examinations, together with those of the weekly quizzes, frequent written tests and the laboratory books, which are corrected and graded every two weeks.

PHARMACOLOGY

The course in Pharmacology is planned to give the student a knowledge of those drugs which will be of the most importance to him in the prophylaxis and therapeutics of disease. The course embodies the essentials of pharmacy, prescription writing, chemistry of drugs, pharmacognosy, toxicology and pharmacology or pharmacodynamics. Pharmacology in its broad sense includes all of those branches, but in its restricted sense it refers to the action of drugs on the body and the actions of the body on the drugs. The course differs from the usual courses given under the name of Materia Medica in that the physiological action of the chemical agents is thoroughly studied by the laboratory method, and receives the bulk of the time allotted to the subject, while the pharmacognosy or the botany of drugs, which in the old course was given much time, is treated in a more limited manner. Three lectures are given per week to the junior class during the first semester. A portion of each lecture period is used for review or recitation upon work considered in previous lectures, and upon the textbook references. Two laboratory periods of three hours each are required for each week. The laboratory work is supplemental to the lectures.

In the earlier part of the course, the essentials of pharmacy, the methods of preparation and dispensing of drugs, the incompatibilities likely to occur in prescribing and a necessary amount of chemistry of drugs are presented to the student, both in the lecture and laboratory.

the Physiological Chemistry laboratory, which is thoroughly fitted up for chemical work.

The laboratory work on the physiological action of drugs consists chiefly of experiments on mammals. Frogs are used whenever they serve the purpose fully. This part of the work is done in the physiological laboratory which is well adapted to this purpose. The students carry on their experiments after the same general methods as were taught in their Physiology of the preceding year. Such experiments are performed as the following: the production of tetanus in frog by strychnine; the effect of caffeine on blood pressure and kidney excretion; the effect of quinine on differentiated protoplasma; the production of diuresis by certain salts as sodium sulphate and sodium citrate; poisoning by morphine and treatment; poisoning by carbolic acid and its treatment; effect of nitro-glycerin, digitalis alcohol etc. on the blood pressure, heart, kidney, etc.

The student thus gains a knowledge of the actions of drugs based upon experimental evidence which enables him to prescribe knowing just what result he can expect. Too much stress cannot be placed upon this, as empirical therapeutics is highly unsatisfying and is doomed to be replaced by an exact and scientific therapeutics.

RESEARCH COURSES
The staff and advanced students in the department will meet fortnightly to discuss the recent literature on physiological subjects and present the results of their own investigations.

Research Courses
7. Research in Physiology (Professors Lyon and Brown)
8. Research in Physiological Chemistry (Professor Neilson)
9. Seminar

Courses in Therapeutics
Remy J. Stoffel, MD, Professor
Hermann H. Born, MD, Professor
1. Therapeutics, Lectures 2 hrs/week to junior class (Professor Stoffel)
2. Demonstrations and Experimental Work in Electrotherapeutics, including the study of electrical phenomena with relation to physiology of the body and the therapeutic application of electricity in all its aspects (Professor Born)

Faculty in the Department during the Lyon Era
Members of the Department and Their Year of Appointment
- Dr. Jacob Friedman, 1903
- Dr. O. H. Brown, appointed instructor, 1904; appointed assistant professor, 1905
- Dr. E. M. Williams, appointed instructor, 1908
- Dr. J. R. Brandon, appointed instructor, 1912
- Dr. L. F. Shackell, appointed instructor, 1910; appointed assistant professor, 1914
- Dr. Charles H. Neilson, appointed associate professor, 1904

The Joseph Era (1913–28)
Biographical Sketch of Don R. Joseph, MD
Dr. Don R. Joseph was the third chair of the Department of Physiology. He served in this capacity for fifteen years, from 1913 to 1928. He also was responsible for Pharmacology from 1914 to 1921. Dr. Joseph was born in 1881 and received a bachelor of science from the University of Chicago in 1904. He received his master of science and doctor of medicine from Saint Louis University in 1906 and 1907, respectively. He was a postdoctoral fellow at the Rockefeller Institute for Medical Research from 1907 to 1908, an assistant professor of physiology from 1908 to 1910, and an associate professor of physiology at

Bryn Mawr College from 1912 to 1913. He accepted the position of professor of physiology and director (chair) of the department at the Saint Louis University School of Medicine in 1913 and served in this capacity until 1928. He also served as vice dean of the medical school from 1919 to 1927 and associate dean from 1927 to 1928. Dr. Joseph was a captain and then a major in the Medical Reserve Corps in 1918 and 1919. His research involved the physiological and toxic effects of ions, as well as nerve irritability and vasomotor centers in shock. Dr. Joseph was active in planning and building Schwitalla Hall, especially the Caroline Street wing and the Grand Avenue front, which extended from Caroline Street to the middle wing. Notes from Dr. Schwitalla, dean of the School of Medicine during this time, state that "he worked hard for the school, skipped his vacations, and took care of a sick member of his family and when he contracted a respiratory infection, it caused his death on July 11, 1928." [6, 7, 20]

Aims of the Department during the Joseph Era

The aims of the Department of Physiology and Pharmacology during the Joseph era were essentially the same as described previously for the Lyon era, minus the parts concerning physiological chemistry, and will not be repeated here. Likewise, the contents of the courses in physiology and pharmacology were similar to those described above.

Therapeutics

In addition to the courses in physiology and pharmacology, there was also a course entitled "Therapeutics" that was taught by Dr. Jacob Friedman and Dr. Hermann Born from 1905 to 1918, when Dr. Friedman died. The course consisted of the following: (6)
1. Lectures in Therapeutics, 2 hrs/week to junior students (Professor Friedman)
2. Clinical Lectures in Medicine with special reference to Applied Therapeutics, 1 hr/week at City Hospital to junior and senior students (Professor Friedman)
3. Lab course in Electrotherapeutics including the study of electrical phenomena with relation to physiology of the body and the therapeutic application of electricity in all its aspects (Professor Born)

Faculty in the Department of Physiology and Pharmacology during the Joseph Era

- Dr. L. F. Shackell, appointed instructor, 1910; appointed assistant professor, 1914
- Dr. H. Whelan, appointed instructor, 1915; appointed assistant professor, 1920
- Dr. C. F. Sutherland, appointed instructor, 1917
- Dr. J. E. Thomas, appointed instructor, 1918; appointed assistant professor, 1920
- Dr. W. W. Hanford, appointed instructor, 1921
- Dr. F. E. Franke, appointed instructor, 1923
- Dr. A. B. Hertzman, appointed assistant professor, 1927 [6, 7]

Faculty during the Lyon and Joseph Eras

(By year of appointment)

Jacob Friedman, MD (1903)

Dr. Friedman earned a degree in chemistry from Washington University in St. Louis in 1876 and an MD from St. Louis Medical College in 1878. He was a physician at St. Louis City and Quarantine Hospital in 1878. He then taught as an associate professor in chemistry at St. Louis Medical College from 1878 to 1892 and as a professor in chemistry at the Beaumont Hospital College of Medicine from 1892 to 1902. He then served as Professor of Therapeutics, Saint Louis University School of Medicine from 1903–18. [6]

Charles H. Neilson, PhD, MD (1904)

Dr. Neilson received a BA from Ohio Wesleyan University in 1894, followed by an MA from that institution in 1897. At the University of Chicago, he was a fellow in physiology in 1900, a research assistant in physiology from 1901 to 1902, and an associate in physiology from 1902 to 1904. He earned a PhD from the University of Chicago in 1903 and an MD from Rush Medical College in 1905. He was assistant city pathologist in St. Louis in 1906 and an associate professor (1904–07) and a professor (1907–11) of physiological chemistry at Saint Louis University. He was also a member of the staff at St. Mary's Infirmary from 1907 to 1908 and was the chief of staff at Alexian Brothers Hospital in 1909. He became professor of medicine and director of the department at Saint Louis University School of Medicine in 1911. [6, 7]

Leon Francis Shackell, PhD (1910)
He served as an assistant in nutritional investigation at the University of Illinois in 1904, an assistant chemist at the University of Illinois in 1904, an assistant chemist at the Missouri Agricultural Experimental Station from 1902 to 1908. He received his BS from Saint Louis University in 1910 and appointed an assistant in physiology at Saint Louis University from 1908 to 1910, a professor of physiology at the dental school and an instructor of physiology at the university from 1910 to 1911, an instructor of physiology at Saint Louis University from 1910 to 1911, an assistant professor in physiology in 1914, and an instructor in pharmacology at the university from 1911 to 1916. He left the university and served as professor of physiology and physiological chemistry at the University of Utah from 1916 to 1925.

John R. E. Brandon, MA (1912)
Mr. Brandon earned his BA from Yale University in 1906 and his MA from Saint Louis University in 1912. He served as an instructor at Spencer College in Jersey City from 1907 to 1908 and as principal of Stevens Institute in New York City from 1909 to 1911. He was appointed instructor in physiology at Saint Louis University in 1912.

Mortimer Bye, BA (1913)
Mr. Bye earned a BA at Johns Hopkins University in 1901 and taught as a lecturer at the College of Physicians and Surgeons in Baltimore from 1902 to 1903 and at the Ohio Mechanics Institute from 1911 to 1912. He was appointed instructor in pharmacology at Saint Louis University in 1913.

Lawrence Schlenker, MD (1913)
Dr. Schlenker earned his PhG from the St. Louis College of Pharmacy in 1906 and his MD from the Saint Louis University School of Medicine in 1910. He interned at the St. Louis City Hospital from 1910 to 1911 and served as an assistant dispensary physician from 1911 to 1913 and as an assistant in pharmacology at Saint Louis University in 1913.

J. L. Shipley, BA (1914)
Mr. Shipley earned his BA from the University of Arkansas in 1909 and was a Rhodes Scholar at the University of Oxford from 1911 to 1914. He was appointed assistant in physiology at Saint Louis University in 1914.

Homer Wheelan, BA (1914)
Mr. Wheelan earned a bachelor's degree from Washington State University in 1911. He served as an assistant in biology at the University of Oregon from 1911 to 1912 and as a lab assistant in physiology at the University of Chicago from 1913 to 1914. He was appointed instructor in physiology at Saint Louis University in 1914.

George Fred Sutherland, PhD (1917)
Dr. Sutherland earned a BA in 1913 and an MA in 1914, both from the University of Illinois. He followed this with a PhD from the University of Chicago in 1917. He was a research assistant in zoology at the University of Illinois from 1913 to 1914 and an assistant in zoology and comparative anatomy at the University of California from 1914 to 1915. He was a fellow in physiology at the University of Chicago from 1915 to 1916 and an assistant in physiology at the same institution from 1916 to 1917. He was appointed instructor in pharmacology at Saint Louis University in 1917.

J. Earl Thomas, MD (1914 and 1921)
Dr. Thomas attended the University of Washington from 1911 to 1913 and received his BS and MD from Saint Louis University in 1918. He was appointed assistant in physiology at Saint Louis University from 1914 to 1918, instructor from 1918 to 1920, and assistant professor in 1920. He served as assistant professor at West Virginia University from 1920 to 1921 and once again as assistant professor at Saint Louis University in 1921. From 1921 to 1956, he was a professor at Jefferson Medical College in Philadelphia. In 1956 he was appointed head of the Department of Physiology at the College of Medical Evangelists in Los Angeles.

Wesley W. Hanford, MS (1921)

Mr. Hanford earned his BS from Wesleyan University in 1913 and his MS at the University of Illinois in 1915. He was appointed instructor in physiology at Saint Louis University in 1921.

Florent E. Franke, MD (1923)

Dr. Franke earned his MD at the Saint Louis University School of Medicine in 1918 and interned at the St. Louis City Hospital from 1918 to 1920. He was appointed instructor in physiology at Saint Louis University in 1923. He was promoted to assistant professor in 1932, associate professor in 1947, and full professor in 1962. He was one of the first scientists to study the effects of smoking and tobacco. Following his retirement in 1963, he served as professor emeritus until his death in 1994 at the age of ninety-eight. He wrote a history of the Department of Physiology at Saint Louis University [7] in 1966 at the time of Dr. Alrick Hertzman's retirement that was very useful to Dr. Westfall in preparing this book. Dr. Franke was a faculty member in physiology at Saint Louis University for a total of forty years. An award for the most outstanding medical student studying physiology, the Franke Award, was named in his honor in 1975.

B. Geret Gosson, BS, MD (1923)

Dr. Gosson earned his BS at Saint Louis University in 1920 and his MD in 1922. He interned at St. Mary's Infirmary from 1922 to 1923. In 1923, he was an assistant physician at the St. Louis City Dispensary and apointed assistant in anatomy, physiology, and medicine at Saint Louis University.

Graduates during the Joseph Era

The first recorded advanced degrees in physiology at Saint Louis University were granted while Dr. Joseph was chair and are listed below.

Master of Science Graduates

- 1927   Hubert Kerper–Anatomy and Physiology
- 1928   Sister Mary Helen Denver–Physiology and Biology
- 1928   Catherine Kahut–Biochemistry and Physiology

Doctor of Philosophy Graduates

- 1927   James Owen Ralls–Biochemistry, Physiology, Anatomy
- 1928   Clarence Jacob Weber–Biochemistry, Bacteriology, Physiology

Chapter Four
The JOHN AUER ERA, DEPARTMENT of PHARMACOLOGY
1921–1948

Biographical Sketch of John Auer, MD

Until the appointment of Dr. John Auer, the chairs of the Department of Physiology (Dr. Shattinger, Dr. Lyon, and Dr. Joseph) also had responsibility for pharmacology. In 1921 Dr. John Auer was appointed professor and chair of the newly established Department of Pharmacology at Saint Louis University School of Medicine. Dr. Auer was born in Rochester, New York, on March 30, 1875. He earned a bachelor's degree at the University of Michigan in 1898, followed by a medical degree from the Johns Hopkins University School of Medicine in 1902. Dr. Auer was a fellow and assistant at the Rockefeller Institute (now Rockefeller University) from 1903 to 1906. He then was an instructor in physiology at Harvard University from 1904 to 1908 before returning to the Rockefeller Institute as an associate and associate pharmacologist, where he remained for thirteen years, from 1908 to 1921.

During and after World War I, from 1917 to 1922, Dr. Auer was a major in the Army Medical Reserve Corps. He then moved to St. Louis and was appointed professor and chair of the newly formed Department of Pharmacology at Saint Louis University School of Medicine, where he served from 1921 to 1948. He was also pharmacologist for the St. Mary's Hospital group.

While at Rockefeller Institute, Dr. Auer began a collaboration with the well-known physiologist Samuel James Meltzer; they worked together for eighteen years and published twenty-five papers together. Dr. Auer also married Melzer's daughter, Clare. Dr. Auer made several important scientific discoveries and provided numerous contributions to the medical literature. He identified rod-shaped crystalline inclusions, present in the cytoplasm of myeloblasts, myelocytes, and monoblasts of a patient with acute leukemia. These so-called "Auer bodies" are named for him. While in the Army Medical Corps, he was the first to apply the relaxing effects of magnesium sulfate to a case of tetanus through intravenous injection. This led to studies on the anesthetic and other effects of magnesium sulfate. He gave the first account of the physiological effects associated with fatal anaphylactic shock in the guinea pig. These studies led to the hypothesis by Meltzer that bronchial asthma may be caused by hypersensitivity to foreign substances. His most important contribution may be that he devised a method of ventilating the lungs with anesthetic gas without lung movements, e.g., during open thorax surgery. This method of intratracheal intubating to give continuous respiration without respiratory movements was used all over the world and marked the beginning of modern endotracheal anesthesia. Other research involved the motor phenomenon of the gastrointestinal tract and the effects of antiseptics and war gas on the function of the blood, bladder, and kidneys.

FIG. 11 Dr. John Auer

Dr. Auer authored over 150 papers in leading journals such as the *American Journal of Physiology*, *Science*, the *Journal of Experimental Medicine*, and the *Journal of Pharmacology and Experimental Therapeutics*. He actively published papers in the scientific literature, literally to the time of his death in 1948. For instance, in 1947 he published seven papers, and in 1946, eight papers. Dr. Auer held memberships in many professional societies, including the American Association for the Advancement of Science, the American Physiological Society, the Association of American Physicians, and the Society for Experimental Biology and Medicine. He was one of the founding fathers of the American Society for Pharmacology and Experimental Therapeutics (ASPET; more on this later). He served terms as councilor of ASPET in 1918 and secretary from 1911 to 1915, as well as three years as president from 1924 to 1927. He was elected to membership of Phi Beta Pi and the Harvey Society.

According to Dr. Alphonse Schwitalla, SJ, dean of the School of Medicine from 1928 to 1948, "Dr. Auer was a well-rounded scholar and read French, German and Latin authors in the original language. He took a great love of art, particularly Matisse, at a time when Matisse's greatness was still not recognized by the establishment. Dr. Auer himself was an amateur painter. In addition, he was a keen gardener and liked to be called a garden master. He loved intellectual argumentation and could be heavily engaged in debates. It was common for him to remain after lunch in order to discuss with younger colleagues. He often took an oppositional stand and advocated inopportune views." [20] Dr. Auer was very interested in medical education and was a popular and dedicated teacher. According to Father Schwitalla, "education was one of his outstanding traits." Schwitalla also stated, "If there is in the graduates of Saint Louis University School of Medicine a trait expressive of a high regard for scientific medicine no less than for the personal services of a physician to his patient, it is due in large measure to John Auer." Dr. Auer died of a coronary thrombosis at St. Mary's Hospital on April 30, 1948.

Following Dr. Auer's death, Father Schwitalla wrote several essays praising his virtues as a scientist, teacher, and humanitarian. He closed one of his commentaries with this statement: "His loss to the School of Medicine is irreparable. ... Every one of our alumni will sorrow with our School of Medicine in this loss." [20]

FIG. 12 Dr. Alphonse M. Schwitalla, SJ

The New Department: Objectives and Aims

Dr. Auer released this statement regarding the new department at the beginning of his tenure as chair:

> With the present academic year the Department of Pharmacology enters the Medical Faculty of the school as a separate independent unit, a position which this branch of science demands in developing competent physicians for the service of the community.
>
> Labs have been provided on the top floor of the new wing under construction. They are adequately equipped for the group study of fundamental pharmacological experiments.
>
> The main object of both the laboratory and didactic instruction will be to provide that scientific basis upon which all rational administration of drugs necessarily rests. Such training should develop the student's judgement so that in active practice he will be able to distinguish with some certainty between rational and irrational therapeutic interventions. The treatment of the subject will therefore stress pharmacology as a medical subject, but without slighting its broader aspects as a branch of biology.
>
> Certain omissions will be made in the course, but these are fully warranted because all textbooks and pharmacopeia carry more or less ballast which is unnecessary for the future medical practitioner. Moreover, it is felt that this reduction of the pharmacological load can be carried out more intelligently by the medically trained instructor than

Synopsis of the Pharmacology Course

The main object of Courses 101 and 102 is to provide the scientific basis upon which the rational use of any therapeutic interference necessarily rests. In order to develop the student's intellectual independence and judgment the laboratory work precedes the didactic course by one semester.

In the laboratory course a series of important pharmacological experiments is carried out by groups of students. These groups are composed of two men; occasionally four work together. A number of groups, generally seven, form a section. Each member of a group alternately acts as group-recorder for an experiment. The group-recorders of a section, in rotation, act as section-recorders. At the end of each week every section-recorder submits a composite graph or tabulation showing the individual results obtained by each group in his section. These composites then are analyzed by the students under guidance at a two-hour conference period. The entire work of the class, therefore, is seen by each student under conditions which permit critical analysis. Selected composites are posted on bulletin boards so that they are available at all times for further study. All experiments in the form of adequate protocols must be recorded by each student and from these protocols inferences warranted by the evidence must be drawn. Great emphasis is laid upon the vital necessity of control experiments. Every animal must be carefully autopsied and the salient findings recorded.

Periodically "unknowns" are issued to each student for the determination and analysis of their more obvious actions by physiologic methods. These tests serve as examinations. No textbook is used in the laboratory course but a laboratory guide is issued to each student which briefly outlines the procedure for each experiment, and which indirectly leads the attention into certain paths. The use of textbooks is discouraged in order to avoid a biased mind. As far as possible students are encouraged to test their explanations of observed phenomena by specific experiments devised by themselves.

The didactic course consists of lectures and informal conferences; no demonstrations are given because they are generally of little value in classes numbering more than 25 students. In this course, pharmacology is treated as a medical subject but without slighting its broader aspects as a branch of Biology. The toxicological action of drugs is considered here as well as in the laboratory course. As the lecture course is designed for medical students, a reduction of the pharmacological load is achieved by the omission of some of the medically less important sections of the pharmacological knowledge. In the didactic course standard textbooks must be used by the student.

In the graduate courses the functional action of drugs is analysed more thoroughly in the laboratory than is possible in Course 101. In general, the work may be restricted to drugs affecting a certain system of the body, or the various systems of the body may be investigated when influenced by a certain drug. A reading knowledge of German and French is desirable for Courses 101 and 102. For Courses 201, 299 a reading knowledge of German is necessary and of French valuable. [6]

101. EXPERIMENTAL PHARMACOLOGY. A laboratory course.

Conferences, two hours a week; laboratory, eight hours a week (four-hour periods); second semester.

Prerequisites: Gross Anatomy, Physiology, Biochemistry, Pathology.

Professors Auer, Shaklee and Assistants

Junior Year

102. PHARMACOLOGY. Lectures and informal conferences. Five hours a week; first semester.

Prerequisites: Course 101 or its equivalent.

Professors Auer, Shaklee and Assistants

Graduate Courses

201. ADVANCED LABORATORY WORK IN PHARMACOLOGY. Chiefly laboratory work with informal conferences and study of the literature.
Prerequisites: Course 101 or its equivalent.
Credit to be arranged.
Professors Auer, Shaklee and Staff

291. JOURNAL CLUB. Meetings held once a week, both semesters. Pharmacology journal club for members of the staff, graduate students in Pharmacology and advanced undergraduates.

STAFF
One hour a week; one credit hour.
Given annually.

299. RESEARCH IN PHARMACOLOGY. Investigation of problems in which the Department is interested.
Professor Auer

---

by the student himself when he lightens his mental burden at the end of a course of instruction. [6]

Formation of the American Society for Pharmacology and Experimental Therapeutics (ASPET) in 1908

On the invitation of Dr. John J. Abel, who was the professor and chair of the Department of Pharmacology at the Johns Hopkins University School of Medicine and was considered the "Father of American Pharmacology," eighteen pharmacologists met at Abel's laboratory at Johns Hopkins to organize a new society. They elected Abel as temporary chairman and Dr. Reid Hunt from the Hygienic Laboratory in Washington, DC, as temporary secretary. Hunt took three pages of minutes, which he and Abel both signed, and had them micrographed.[21] Four articles of agreement were unanimously adopted. They are as follows:

> 1. *In order to further the growth of pharmacology and experimental therapeutics in this country and to facilitate personal intercourse among investigators in these branches of science, we hereby organize the American Society for Pharmacology and Experimental Therapeutics (ASPET) and subscribe ourselves thereto as its founders.*
>
> 2. *The management of the Society will be left in a Council of seven members—a President, a Secretary, a Treasurer and four councilors.*
>
> 3. *The Council is to prepare a constitution to consider ways and means for permanent establishment of the Society, and furtherance of its purpose by calling meetings.*
>
> 4. *Twelve members in person or by proxy will constitute a quorum until a constitution is adopted.*

Elected as officers for the following year were J. J. Abel as president, R. Hunt as secretary, A. S. Leevenhart as treasurer, and S. J. Meltzer, T. Sollman, C. W. Edmunds, and A. C. Crawford as councilors. Before adjournment, Abel also announced the establishment of the Journal of Pharmacology and Experimental Therapeutics, which has become the gold standard for pharmacology papers over the years. As can be seen from the following list, John Auer, who was at the Rockefeller Institute at the time, was among the eighteen founding fathers.

Founders of the American Society for Experimental Therapeutics

- John J. Abel, Johns Hopkins University School of Medicine, Baltimore, MD
- Carl L. Alsberg, Bureau of Plant Industry, Department of Agriculture, Washington, DC
- John Auer, Rockefeller Institute, New York City, NY
- Albert C. Crawford, Bureau of Animal Industry, Department of Agriculture, Washington, DC
- Charles W. Edmunds, University of Michigan, Ann Arbor, MI
- J. A. English Eyster, University of Virginia, Charlottesville, VA
- W. Worth Hale, Hygienic Laboratory, Washington, DC

- Robert A. Hatcher, Cornell Medical College, New York City, NY
- Velyien E. Henderson, Pharmacological Department, University of Toronto, Toronto, Canada
- Reid Hunt, Hygienic Laboratory, Washington, DC
- Arthur S. Leevenhart, University of Wisconsin, Madison, WI
- Samuel A. Matthews, University of Chicago, Chicago, IL
- Samuel J. Meltzer, 13 W. 121st Street, New York City, NY
- William Salant, Bureau of Chemistry, Department of Agriculture, Washington, DC
- Torald Sollman, Medical Department, Western Reserve University, Cleveland, OH
- Maurice V. Tyrode, Harvard Medical School, Boston, MA
- Carl Voegtlin, Johns Hopkins Medical School, Baltimore, MD
- Horatio C. Wood Jr., University of Pennsylvania, Philadelphia, PA

Fifteenth Annual Meeting of ASPET Comes to Saint Louis University in 1923

During the time Dr. Auer was chair of the department, he was able to attract a meeting of ASPET to St. Louis under the sponsorship of Saint Louis University and Washington University. The first two days were held at Washington University and the last day at Saint Louis University. John Auer was elected president at this meeting and presided at the ASPET meetings held in 1924, 1925, and 1927. According to Dr. K. K. Chen, historian for ASPET: "Those who heard [John Auer] speak could not help admiring his eloquence and smoothness. He conducted Society business with the greatest precision." [21] Other highlights of the Fifteenth Annual Meeting were that clinical pharmacology remained with ASPET, as evidenced by L. G. Rowntree's paper on glandular therapy in Addison's disease, and H. H. Young's laboratory and clinical experience with certain germicides. Business matters were as usual: election of officers and council members; election of five new members to ASPET, and the nomination of R. Hunt to the membership committee. The last session drew a large audience, with 132 members and 133 guests. [21]

The John Auer Lectureship

In recognition of his outstanding contributions to pharmacology as a research worker and teacher in his specialty, the Phi Beta Pi Medical Fraternity at Saint Louis University presented an annual lectureship to the university named in honor of Dr. John Auer, professor of pharmacology and director of his department. The lectureships existed from 1944 to 1967 and featured twenty-one distinguished individuals as lecturers. [6] It is unclear why the lectureship was not continued after 1967. The lecturers were:

- 1944 Dr. Warfield T. Longcope, Professor of Medicine, Johns Hopkins University
- 1945 Dr. Walter C. Alvarez, Professor of Medicine, University of Minnesota
- Spring 1947 Dr. Owen H. Wangensteen, Director of the Department of Surgery and Surgeon-in-Chief of University Hospital, University of Minnesota
- Fall 1947 Dr. Reginald H. Smithwick, Professor of Surgery, Boston University School of Medicine
- 1948 Dr. Emil Novak, Assistant Professor of Gynecology, Johns Hopkins University School of Medicine
- 1949 Dr. John Peters, Professor of Medicine, Yale University School of Medicine
- 1950 Dr. Joseph Johnston, Pediatrician-in-Chief, Henry Ford Hospital, Detroit
- 1951 Dr. Richard B. Cattell, Surgeon, Lahey Clinic; Surgeon-in-Chief, New England Deaconess Hospital, Boston
- 1955 Dr. Edward T. Beattie Jr., Chairman, Department of Surgery, Presbyterian Hospital, Chicago
- 1956 Dr. William A. Sodeman, Professor of Medicine, University of Missouri School of Medicine
- 1957 Dr. Mark M. Ravitch, Associate Professor of Surgery, Johns Hopkins University School of Medicine
- 1958 Dr. Carleton B. Chapman, Professor of Medicine, Texas Southwestern Medical College

- 1959 Dr. Alexander G. Gutman, Professor of Medicine, Columbia University College of Physicians and Surgeons
- 1960 Dr. Daniel C. Darrow, Professor of Pediatrics, University of Kansas School of Medicine
- 1961 Dr. James G. Hilton, Assistant Professor of Medicine, Columbus University, College of Physicians and Surgeons
- 1962 Dr. Richard Varco, Professor of Surgery, University of Minnesota
- 1963 Dr. Alvin K. Merendino, Professor of Surgery, University of Washington in Seattle
- 1964 Dr. William C. Thomas Jr., Professor of Medicine and Director of the Department of Medicine, University of Florida
- 1965 Dr. Robert M. Kark, Professor of Medicine, University of Illinois
- 1966 Dr. Hans Popper, Professor and Chief of Pathology and Academic Dean of Mt. Sinai School of Medicine
- 1967 Dr. Rene Menguy, Professor and Chairman of the Department of Surgery, University of Chicago

Faculty during the Auer Era

(By year of appointment)

Alfred Ogle Shaklee, MD (1921)

Dr. Shaklee earned his BS at the University of Chicago in 1899 and his MD from Saint Louis University School of Medicine in 1908. He did medical research at the Rockefeller Institute from 1908 to 1910 and was head and associate professor of pharmacology at the University of the Philippines from 1910 to 1913. He was assistant professor of pharmacology at the University of Illinois from 1914 to 1916, professor of physiology and biochemistry at Fordham University from 1916 to 1918, and associate professor of physiology and pharmacology at the University of Texas from 1918 to 1921. He was appointed associate professor of pharmacology at Saint Louis University in 1921 and served the department for thirty years, until his death in 1951 at the age of eighty-five.

Frank Oscar Anderson, MD (1922)

Dr. Anderson earned a BA from the University of Missouri in 1921 and an MD at Saint Louis University in 1924. He was an assistant in pharmacology at Saint Louis University in 1922 and an instructor in 1924. He interned at Mullanphy Hospital from 1924 to 1925.

James Boswell Mitchell, PhD (1930)

Dr. Mitchell earned his BS and MS at Emory University in 1923. These were followed by a PhD at the University of Chicago in 1928. He served as instructor in pharmacology at Saint Louis University School of Medicine from 1930 to 1933.

Lloyd D. Seager, MD (1933)

Dr. Seager earned an BA from Milton College and was an assistant in physiology at the University of Illinois from 1927 to 1929. He earned an MD at Illinois in 1934. He served as an instructor in pharmacology at the Saint Louis University School of Medicine from 1933 to 1937.

Hugo M. Krueger, PhD (1939)

Dr. Krueger received a BA from the University of Denver in 1924, followed by an MA from that same institution in 1926. He was an instructor in Latin at the University of Denver from 1924 to 1927 and an assistant in chemistry from 1926 to 1927. He was an assistant in physiology at the University of Michigan from 1927 to 1930 and earned his PhD from that school in 1930. At the University of Michigan, he was a research instructor in pharmacology from 1931 to 1933, a research associate in pharmacology from 1933 to 1934, and an assistant professor from 1933 to 1939. He was appointed senior instructor in pharmacology at Saint Louis University in 1939, senior assistant professor from 1941 to 1944, and associate professor in 1944.

Benjamin DeBoer, PhD (1946)

Dr. DeBoer earned a BA from Calvin College in 1933, followed by an MA in 1938 and a PhD in 1942, the latter two from the University of Missouri. He served as an instructor in physiology and pharmacology at the University of Missouri from 1941 to 1943 and an assistant professor from 1943 to 1946. He was appointed assistant professor of pharmacology at Saint Louis University in 1946.

Chapter Five
DEPARTMENT of PHARMACOLOGY:
The ROBERT H. K. FOSTER, ERWIN NELSON, & ELIJAH ADAMS ERAS
1948–1963

The Robert H. K. Foster Era (1948–52)

Upon the death of John Auer, a national search was conducted to find a replacement. Under the leadership of Edward Doisy, professor and chair of the Department of Biochemistry, and Alrick Hertzman, professor and chair of the Department of Physiology, letters were sent to several prominent individuals seeking the names of suitable candidates. Among those consulted were Dr. K. K. Marshall of Johns Hopkins University, Dr. Carl Schmidt of the University of Pennsylvania, Dr. Granville A. Bennett of the University of Illinois, and Dr. E. M. K. Gerling of the University of Chicago. Numerous outstanding candidates were identified, including Leon H. Schmidt, the director of the Institute of Medical Research at Christ Hospital in Cincinnati; A. Calvin Bratton, the director of research at Park Davis and Co.; James A. Shannon, the director of the Squibb Institute for Medical Research; Seymour S. Kety of the University of Pennsylvania; Karl H. Beyer, director of pharmacological research at Sharpe and Dohme Laboratories; Robert Woodbury of the University of Tennessee; Benedict Abreau of the University of California; Maynard Chenoweth of the University of Michigan; George Sayers of the University of Utah; Alfred Gilman of Columbia University; Klau Unna of the University of Illinois; Julius Coon of the University of Chicago; and R. H. K. Foster of Saint Louis University. Most of these candidates later assumed important leadership roles as chairs of academic departments or directors at the National Institutes of Health (NIH) or universities and pharmaceutical companies throughout the United States.

Biographical Sketch of Robert H. K. Foster, MD, PhD

From this outstanding list of candidates Dr. Robert H. K. Foster was chosen and appointed chair of the Department of Pharmacology in 1948 to replace John Auer. Dr. Foster served as chair until 1952.

Dr. Foster earned a bachelor of chemical engineering from Ohio State University in 1923. He followed this with a doctor of philosophy from the University of Chicago in 1932 and a doctor of medicine from Rush Medical College in 1935. He had served as a graduate assistant in pharmacology at the University of Chicago from 1931 to 1932 and an instructor in 1932. He completed an internship at the Ravenswood Hospital in Chicago from 1934 to 1935. He was appointed associate professor in the Department of Pharmacology at the Saint Louis University School of Medicine in 1945 and director of the department in 1948. Dr. Foster served as chair of the Department of Pharmacology at Saint Louis University from 1948 until 1952. He would often give refresher courses around the country, such as "Pharmacology and the Use of Narcotics," which he gave at a meeting of the Chiropodist Society of Texas in December 1951. He resigned on July 31, 1952, and went into the private practice of medicine.

Synopsis of the Pharmacology Course

Since Dr. Foster served as an associate professor under the chairmanship of John Auer for several years, he was familiar with the objectives of the laboratory and didactic pharmacology course presented during the Auer era. The course was continued as such and the details as described under Dr. Auer's era will not be repeated here.

Graduate Program

Although there were no graduate degrees given while Dr. Foster was chair of the Department of Pharmacology, he did advertise teaching fellowships in pharmacology with the opportunity to work toward a master's degree or doctorate. These advertisements stated:

The stipend for the teaching fellowships would be $1,200.00 annually plus graduate tuition fees. These advertisements are stated below in the blue box.

The time would be divided between graduate courses, research and teaching. These three categories will be integrated to provide the most efficient training for the candidate and to fulfill all of the necessary requirements in the Graduate School.

Prerequisites for the admission of a candidate besides those of the graduate school are two years of medicine or the approximate equivalent. The applicant must have completed the usual medical course in pharmacology as well and biochemistry and physiology. An exception exists in the case of pharmacology when, as at this school, the lecture course is normally not given until the Junior year of medicine. In addition applicants who have more than the premedical requirements in chemistry, especially organic, and biology, will be given preference, other conditions being equal.

Applicants who have completed their medical training and have received their MD degree and who would be interested in a career in pharmacology will be given special consideration. Applicants who have only partially completed medical studies and wish to obtain their MDs as well as a degree in pharmacology will also be given special consideration. Arrangements should be made to enable such men to secure both degrees.

There is perhaps no other field of medicine in which as great a dearth of trained men exists as in pharmacology. There are at least three reasons for this: One, the demand for pharmacologists in industry. Two, the time at which candidates are best secured, that is, after two years of medicine, is a stage when most potential candidates are very anxious to complete medical studies. Three, there has never been enough graduate training in pharmacology to supply the demand. Many schools have never provided the training and hence we see pharmacologists obtained by the metamorphosis of biochemists or physiologists.

Faculty during the Foster Era

(By year of appointment)

Kazuo K. Kimura, PhD (1949)

Dr. Kimura earned his MS at the University of Nebraska in 1944 and his MD at Saint Louis University in 1953. He was an instructor in physiology and pharmacology at the University of Nebraska from 1944 to 1945 and a Roche Fellow in Pharmacology at the University of Illinois for 1946 to 1949. He was appointed instructor of pharmacology at Saint Louis University in 1949 and promoted to assistant professor in 1952. He resigned July 1, 1954.

Irving A. Coret, BA, MD (1950)

Dr. Coret earned a BA in chemistry in 1940 and an MD in 1943, both from Emory University. He followed this with a residency at Piedmont Hospital in Georgia and a postdoctoral fellowship in the Department of Zoology at the University of Pennsylvania, working with Dr. L. V. Heilbrunn from 1948 to 1950. He accepted a position in the Department of Pharmacology at the Saint Louis University School of Medicine in 1950 and had the distinction of serving under five different chairs: Dr. Robert Foster, Dr. Erwin Nelson, Dr. Elijah Adams, Dr. René Wégria, and finally Dr.

FIG. 13 Dr. Irving Coret

Thomas C. Westfall. Dr. Coret himself was executive secretary (equivalent to interim chair) of the department from 1957 to 1958. Dr. Coret's research was in the mathematical analysis of drug receptor interactions and theoretical pharmacology. He was a faculty member in the Department of Pharmacology at Saint Louis University for an astonishing forty years and provided great service through his teaching and committee memberships. Dr. Coret retired in 1990 and served as professor emeritus until his death in 1996.

Benjamin DeBoer, PhD
See page 28.

Alfred Ogle Shaklee, MD
See page 28.

The Erwin E. Nelson Era (1952–57)

Biographical Sketch of Erwin E. Nelson, PhD, MD

Dr. Nelson was born in Springfield, Missouri, in 1891. He graduated with a BS from Drury College in 1914 and an MA in 1916. He followed this with a PhD in physiology and pharmacology in 1920 from the University of Missouri. He was appointed assistant professor of pharmacology at the University of Michigan Medical School, where he studied medicine and graduated with an MD in 1926. He was promoted to associate professor in 1927 and to professor in 1936. He was appointed head of the Department of Pharmacology at Tulane University in 1937, where he remained as professor and head until 1943. In that year Dr. Nelson was appointed director of research at Burroughs, Wellcome & Co. in New York, where he stayed for three years. After that, he worked as chief at the Food and Drug Administration from 1947 to 1950 and as medical director from 1950 to 1952. He was appointed professor and chair of the Department of Pharmacology at the Saint Louis University School of Medicine in 1952 and served until 1957. He retired in Albuquerque, New Mexico, and died in 1966 at the age of seventy-six. Dr. Nelson was noted for his work on the bioassay of ergot, posterior pituitary extracts, and analgesics. He was a member of the American Physiological Society, the Society for Experimental Biology and Medicine, the American Medical Association, the Revision Committee of United States Pharmacopeia (1930–47) and the Committee on Addiction and Narcotics of the National Research Council (1948–58). He also served on the committee writing pharmacology examination questions for the National Board of Medical Examiners. Dr. Nelson served as treasurer for the American Society for Pharmacology and Experimental Therapeutics from 1939 until 1943, when he became the second member of the SLU Department of Pharmacology chosen to serve as president of that society. His term as president expired in 1946, when, with World War II over, he was finally able to preside over the annual meeting held in Atlantic City. It is said that he steered ASPET to almost prewar normalcy. [6, 21]

FIG. 14 Dr. Erwin E. Nelson

The Role of Pharmacology in Medical Education

The following is a written statement about the role of pharmacology in the medical curriculum prepared by Dr. Nelson for James W. Colbert. Dr. Colbert was Dean of the School of Medicine from 1953–61 and incidently was the father of Stephen Colbert, host of *The Late Show with Stephen Colbert* on CBS. The statement was dated May 11, 1954 and is as true today as it was then.

> *The actual role of Pharmacology in medical education is fairly clearly defined by the place given in the sequence of subjects in the medical curriculum of most schools. It follows the courses in which the normal functioning of the organism (man) has been covered in courses in physiology*

and physiological chemistry. It usually follows the courses in bacteriology which today covers not only morphological characteristics but also the physiology and physiological chemistry of these small organisms. Pharmacology is usually contemporary with pathology in most curricula. Pharmacology may be defined as being concerned not only with the morphological changes, but also to an increasing extent with physiological changes induced by the presence of microorganisms, by failure of proper development, by inadequate nutrition, by imbalance between homeostatic mechanisms, etc. Thinking in the same terms, pharmacology takes for its field the way in which normal or disturbed function can be modified by chemical agents. It is thus very close to the physiological sciences on the one hand, and on the other to therapeutics (that form of therapeutics which involves drugs).

Pharmacology faces a problem which is to some degree unique in the medical curriculum. The chemicals used today as drugs, and the ones he must discuss with his sophomore medical students, are not those in use two or five or ten years ago, and very probably not those which the student himself will use when he gets to his internship. In no area has change been so rapid or extensive in recent years, and there is no reason for believing that it will not continue. Is it possible to teach principles rather than details?

Related to this situation is the fact that the pharmaceutical industry, which inevitably is involved in every advance in drug therapy, is first and foremost a profit-making organization. Of course, it is only by such profits that it can continue the operation of its research and development laboratories. This gives rise to a circular process not unfairly labeled vicious. Nevertheless, when the student leaves school and enters the practice of medicine, there is no area where he will be put under pressure so steadily and probably so effectively as in that having to do with the choice of drugs. It is imperative, therefore that the medical student be given as much as possible to help him make his selections objectively, to give him principles and standards against which to measure poor or inadequate therapeutic studies.

According to Dr. Nelson, his contact with other teachers through his membership on the committee preparing examination questions for the National Board of Medical Examiners led him to believe that the information presented in the courses at Saint Louis University School of Medicine compared very favorably with that presented at other schools.

Faculty Research Interests

As stated by Dr. Nelson, "There is no general program of research into which each staff member must fit. Each man is encouraged to follow his own bent." Areas of research carried out by faculty during the Nelson era are summarized below:

- **Kazuo K. Kimura, MD, Saint Louis University**
  Pharmacology of curare-type compounds.

- **Paul M. Lish, PhD, Saint Louis University**
  Pharmacological and physiological actions of drugs acting at the neuromuscular junction.

- **Irving Coret, MD, Emory University**
  Effects of chemical substances on light production by luminous bacteria and relation of such phenomics to effects of chemical agents on enzymes in general.

- **Paul Bettonville, PhD, University of Illinois**
  Possible role of folic acid in the incidence of chemical carcinoma in mice.

- **Leonard Procita, PhD, University of Michigan**
  Pharmacology of muscle.

- **Erwin Nelson, MD, University of Michigan; PhD, University of Missouri**
  Bioassay of ergot, posterior pituitary extracts, and analgesics.

Graduates during the Nelson Era

Doctor of Philosophy Graduates
- 1955 Paul M. Lish–PhD
- 1957 Ira William Hillyard–PhD

Master of Science Graduates
- 1952 Eaden Francis Keith MS(R)
- 1955 John Walter Kissel–MS(R)

An example of the advertisement to recruit graduate students during the Nelson era is shown on page 33.

Opportunity To Work For Ph. D. In One Of

Basic Medical Sciences

Teaching Assistantship In
Department Of Pharmacology
St. Louis University School Of Medicine

Inquiries are invited from seniors, in liberal arts colleges and pharmacy schools, who are interested in research and teaching in the physiological sciences.

Teaching duties limited almost entirely to the 6-month period from September to February. The remainder of the year is available for study and research.

Initial stipend is $1,500 per annum.

Successful applicant, in addition to assisting in laboratory teaching in pharmacology, will begin program of studies and research leading to Ph.D. in pharmacology.

Interested persons should communicate as soon as possible with:

DR. ERWIN E. NELSON, DIRECTOR
DEPARTMENT OF PHARMACOLOGY
SAINT LOUIS UNIVERSITY SCHOOL OF MEDICINE
ST. LOUIS 4, MISSOURI

Faculty during the Nelson Era
(By year of appointment)

Paul Lish, BS, MS, PhD (1951)
Dr. Lish earned a BS from the Idaho State College of Pharmacy in 1949, an MS from the University of Nebraska in 1951, and a PhD from Saint Louis University in 1955. He served as a teaching assistant at the University of Nebraska from 1949 to 1951, a graduate fellow in pharmacology at Saint Louis University from 1951 to 1952, and a fellow of the American Foundation for Pharmaceutical Education in 1953. His research studied the pharmacological and physiological actions of drugs acting at the neuromuscular junction.

Ira W. H. Hillyard, PhD (1953)
Dr. Hillyard received his BS from Idaho State College in 1949 and an MS from the University of Nebraska in 1951. He served as a graduate assistant at the University of Nebraska from 1950 to 1951 and as an assistant in pharmacology at Saint Louis University from 1953 to 1957. He earned his PhD in pharmacology/physiology in 1957 under the tutelage of Dr. Erwin Nelson and served on the faculty for one year.

Leonard Procita, PhD (1954)
Dr. Procita received a BA from the University of Michigan in 1948 followed by an MS in 1951 and a PhD from the same institution in 1954. He worked at the University of Michigan as a research assistant from 1948 to 1951, a teaching fellow from 1951 to 1952, and a research assistant from 1952 to 1954. He was appointed instructor in pharmacology at Saint Louis University in 1954 and served until 1956.

Ernest C. Griesemer, BS, MS, PhD (1956)
Dr. Griesemer earned his BS from Loyola University in 1950, followed by an MS in 1952, also from Loyola. He received a PhD from Northwestern University in 1955. He served as an instructor in pharmacology at the Stritch School of Medicine from 1950 to 1952 and at Northwestern University from 1952 to 1956. He was appointed instructor in pharmacology at Saint Louis University in 1956.

Kazuo K. Kimura, PhD
See page 30.

Irving A. Coret, MD
See page 30.

The Elijah Adams Era (1958–63)
Biographical Sketch of Elijah Adams, MD

The fourth chair of the independent Department of Pharmacology and the seventh overall was Dr. Elijah Adams, who followed Dr. Erwin Nelson in 1958. Dr. Adams served as professor and chair for five years, from 1958 to 1963. He earned a BA from Johns Hopkins University in 1938 and an MD from the University of Rochester School of Medicine in 1942. He did an internship in medicine at Strong Memorial

FIG. 15 The Department of Pharmacology in 1958. Dr. Adams is in the center, 7th from the right, front row.

Hospital 1942 and 1943 and followed that with a residency in pathology at Grace-New Haven Hospital of Yale University from 1946 to 1947. Dr. Adams was assistant resident in medicine at Grace-New Haven from 1948 to 1949. He was also an NIH research fellow in biochemistry at Yale University from 1947 to 1948. He served as an American Cancer Society fellow in biochemistry at the University of California from 1949 to 1950. Dr. Adams served as associate professor of pharmacology at New York University School of Medicine from 1955 to 1958 prior to accepting the position of professor and chair of the Department of Pharmacology at Saint Louis University School of Medicine in 1958. Dr. Adams's research was in the area of cancer pharmacology, and he served on numerous local and national committees. In 1963, Dr. Adams left the university to accept a position at the University of Maryland.

Synopsis of the Pharmacology Course

The medical pharmacology course and its objectives were essentially the same as that described on page 31 by Dr. Nelson to Dean Colbert in 1954. It consisted of a lecture course covering the important drugs used in therapy. In the course, the major groups of drugs used in medicine were covered, with emphasis on physiological and biochemical principles concerned with the action, absorption, metabolism, excretion, and toxicity of drugs.

Lectures and conferences occupied six hours out of each week, with laboratory work occupying another six hours. Prerequisites included AN131, BCH102, and PY101.

Graduate Courses
Topics in Molecular Biology (given jointly with the Department of Microbiology). The topics course occupied two credit hours per semester.

Journal Club
One hour per week

Research Credit
Students in the PhD program during the Adams era were actually awarded their PhDs during the time Dr. Wégria was chair and are listed in the following chapter on page 41. A unique feature of the medical pharmacology course introduced in 1958 was a teaching experiment involving a series of student reports on selected topics in pharmacology, with special emphasis on recent research developments. It replaced some of the laboratory experiments conducted by students in the past. The following is a description of this exercise as presented to the students in October 1958.

Pharmacology Student Reports

Our motives in this project are manifold: First, we want you each to develop some specific knowledge—in more than textbook detail—of a specific facet of one of the many topics our course touches on. Second, we want you to see and evaluate a sample of the data and interpretations of the original literature, on which monographs and textbook accounts are based. Third, we want to encourage your acquaintance with the library techniques of looking up a narrow subject in depth. Fourth, we want you to teach each other by reporting on many interesting topics more fully than lecture coverage permits. Finally, we want you to attempt the oral exposition of a scientific topic, a skill you may be frequently called on to exercise during your professional life. In the present exercise, the clear and concise presentation of unfamiliar and often complex material to your fellow students is a particularly challenging problem in communication.

The plan of the exercise is to have each laboratory group of four students pick a single topic to report on. The topic may be selected from the large and varied accompanying list or may be your own (in the latter case have it approved by one of the staff). Next sign up for your topic on the sheet provided on the bulletin board.

For convenience, the class has been divided into three teams; I, II and III, each including nine laboratory groups. Each team will meet as a unit and will listen to presentations of the groups within it. The schedule for the sessions of the three teams and the schedule for presentations by the individual groups are shown on the bulletin board.

Because the students in each team will comprise the audience for all topics within that team, it is not possible to have duplication of topics within a team: therefore, the choice of a given topic goes to the first group in that team signing up for it. Duplication of topics in different *teams is permissible.*

The exercise is designed that all individuals have a chance to speak. Each of 4 students in a group should take 10 minutes for an individual presentation within the general topic of the group; this will mean about one hour per topic, including discussion.

We do not want you to summarize review articles, but instead to try to concentrate on one of a few critical papers, with attention to data and interpretations, whenever possible. The students are not to worry about incomplete coverage of the topic they have chosen. It is preferable to try to select and present individual papers.

The exact division of labor in presenting a subject is up to you and will be determined in part by the scope and nature of the topic. With many topics it will be profitable to have the first speaker present historical or general background material, so that the significance of the topic is clear, and the later speakers can dig into details with an oriented audience.

To provide you with some record of the reports you will be hearing, we ask each group to provide a summary of their presentation including references to the papers they report on. The abstract and references together should not exceed one sheet of single-spaced typescript and should reach the department secretary, Mrs. Franke, by noon of the day before your report. The mimeographed abstracts will then be available to the audience at the time of the report.

General references to your topic can probably be found in the bibliographies of your two major textbooks, Drill and Sollman. Please note that Drill's Pharmacology in Medicine is three years newer than Sollman's Manual of Pharmacology.

The list of fifty possible topics provided to the students is shown in Table 3. It is of interest that this exercise was very similar to that conducted in the medical pharmacology course during the Westfall era some twenty-five years later.

Composition of Teams

Team I	Team II	Team III
Section A: Groups 1–7	Section B: Groups 4–7	Section C: Groups 6,7
Section B: Groups 1, 2	Section C: Groups 1–5	Section D: Groups 1–7

Schedule for Teams

	Tues	Thurs
Nov. 13		I
Nov. 18	II	
Nov. 20		III
Nov. 25	I	
Nov. 27		Holiday
Dec. 2	II	
Dec. 4		III
Dec. 9	I	
Dec. 11		II
Dec. 16	III	

Faculty during the Adams Era

(By year of appointment)

Clark Bublitz, PhD (1959)

Dr. Bublitz earned a PhB from the University of Chicago in 1949 and a PhD from that same institution in 1955. He was an assistant in biochemistry at the University of Chicago from 1952 to 1953 and a predoctoral fellow of the Nutrition Foundation from 1954 to 1955. He served as a postdoctoral fellow with the American Cancer Society from 1955 to 1957 and a fellow in physiological chemistry at Johns Hopkins University from 1957 to 1959. He was appointed senior instructor in pharmacology at Saint Louis University in 1959.

Harold L. Segal, PhD (1959)

Dr. Segal earned his BS from the Carnegie Institute of Technology in 1947 and an MS and a PhD from the University of Minnesota in 1950 and 1952, respectively. He served as a research associate at the University of Minnesota from 1949 to 1950. He was a research associate in zoology at UCLA from 1952 to 1954 and an assistant professor in biochemistry at the University of Pittsburgh from 1954 to 1959. He was appointed assistant professor of pharmacology at Saint Louis University in 1959 and was promoted to associate professor in 1960. He served under the chairmanship of both Adams and Wégria. He left the

TABLE 3 Suggested Topics for Student Reports, October 1958

1. Mechanism of morphine tolerance.
2. Mechanism of action of disulfiram (in particular, evidence for inhibition of acetaldehyde dehydrogenase and for acetaldehyde accumulation?).
3. Distribution and metabolism of thiopental in relation to its duration of action. (Question of uptake by adipose tissue and subsequent redistribution.)
4. Mechanism of acetazoleamide diuresis.
5. Mechanism of beryllium poisoning.
6. Experimental ototoxicity of streptomycin and related compounds.
7. Metabolism of C^{14} digitalis.
8. Studies with cardiac catheterization on digitalis action.
9. Mechanism of hypoglycemic action of insulin: current concepts.
10. Human carcinoid tumors: pharmacologic aspects.
11. Storage and release of histamine from mast cells.
12. Relation of thyroid hormone to oxidative phosphorylation.
13. Physiologic defect in myasthenia gravis.
14. Site of attachment to enzymes of diisopropyl fluorophosphate.
15. Mechanism of emetic action of morphine.
16. Digitalis action on ion transport in cells.
17. Mode of action of thiocyanate and other goitrogens.
18. Antagonism between lysergic acid diethylamide and 5-hydroxytryptamine: significance in brain function.
19. Mechanism of the potentiating agent SKF-525A.
20. Direct muscular action of neostigmine.
21. Significance of epinephrine release after adrenergic nerve stimulation.
22. Mechanism of action of penicillin: current concepts.
23. Pyridine aldoxime methiodide as a rational antidote for the phosphate anticholinesterases.
24. Mechanism and locus of dibenamine action.
25. Current concepts of the myoneural blocking action of decamethonium and succinylcholine.
26. Does epinephrine show direct vasodilator action in certain vascular beds?
27. Reserpine interrelations with 5-hydroxytryptamine.
28. Mechanism of the vagal effect of cardiac glycosides.
29. Tolerance to the barbiturates.
30. The CNS actions of curare.
31. Salicylates and connective tissue metabolism.
32. Drug actions on cerebral circulation: recent studies.
33. Drug actions on coronary circulation: recent studies.
34. Carcinogenic effect of croton oil and other irritants.
35. Relation of the reticular activating system to anesthesia.
36. Studies of cardiac muscle using intracellular electrodes.
37. Does "analgesic" activity of drugs depend on "pain threshold"?
38. Hypoxia in induction of N_2O anesthesia: controversial aspects.
39. Nature and pharmacological action of botulinus toxin.
40. Possible bases for insulin resistance in diabetes mellitus.
41. Effects of adrenal cortical steroids on connective tissue growth.
42. Pharmacology of the convulsant barbiturates.
43. The renin-angiotonin system in human hypertension.
44. Pharmacological basis of diagnostic tests for pheochromocytoma.
45. Metabolism of epinephrine: newer findings.
46. Colchicine effects of uric acid metabolism.
47. Mechanism of the antihypertensive action of reserpine.
48. Newer humeral mediators postulated to regulate intestinal motility (Darmstoff, P substance).
49. Mechanism of chlorothiazide action (diuretic and/or hypertensive).
50. Tolbutamide: mechanism of hypoglycemic action.

university in the early 1960s to become professor and chairman of the Department of Biology at the State University of New York in Buffalo.

K. K. Govind Menon, PhD (1960)

Dr. Menon earned a BS from the Maharaja's College in Ernakulam, India, in 1947, followed by an MSc from that institution in 1953. He earned a PhD from the University of Toronto in 1956. He was an assistant in biochemistry and endocrinology at the Banting and Best Department of Medical Research, University of Toronto, from 1953 to 1956. He followed with a postdoctoral position at the University of Pittsburgh in 1956 and 1957 and at Western Reserve University from 1957 to 1959. He was at the Patel Chest Research Institute in India from 1959 to 1960. Dr. Menon was appointed senior instructor in pharmacology at Saint Louis University in 1960.

John C. Loper, PhD (1960)

Dr. Loper earned a BS from Western Maryland College in 1952, an MS from Emory University in 1953, and a PhD from Johns Hopkins University in 1960. He was appointed instructor in pharmacology at Saint Louis University in 1960 and served until 1963, when he retired from the university.

Audrey L. Stevens, PhD (1961)

Dr. Stevens earned her BS from Iowa State University in 1953 and her PhD from Western Reserve University in 1958. She was a National Science Foundation (NSF) postdoctoral fellow in biochemistry at the NIH from 1958 to 1960 and was appointed to assistant professor at Saint Louis University (or SLU) in 1961. Dr. Stevens served until her retirement from the department in 1963.

Seymour Pomerantz, PhD (1961)

Dr. Pomerantz received a BA from Rice Institute in 1948 and her PhD from the University of Texas in 1952. He was appointed assistant professor in pharmacology at Saint Louis University in 1961 and served until 1963, when he left the university.

Irving A. Coret, MD

See page 30.

Biographical Sketch of René William Edward Wégria, MD, MedScD

Dr. René Wégria served as the fifth chair of the independent Department of Pharmacology and the eighth overall at the Saint Louis University School of Medicine from 1963 to 1979. He was born in Fumal, Belgium, on June 9, 1911. He earned his doctor of medicine magna cum laude in 1936 from the University of Liège in Belgium and was Lauréat du Concours Interuniversitaire de Belgique from 1934 to 1936. He was a fellow of the Belgian American Educational Foundation at Vanderbilt University, the Mayo Foundation, and Western Reserve University from 1937 to 1939. Dr. Wégria followed this as a fellow in the Department of Physiology at Western Reserve University from 1939 to 1940 and instructor in physiology at the same institution from 1940 to 1942. From 1942 to 1943, he served as assistant resident in medicine at Cleveland City Hospital in Ohio.

FIG. 16 Dr. Budh Bhagat, Dr. René Wégria (chairman), and Dr. H. H. Oei

From 1943 to 1946, Dr. Wégria was assistant resident in cardiology and assistant in medicine at Presbyterian Hospital and Columbia University College of Physicians and Surgeons in New York City. While at Columbia, he also was awarded a doctorate in medical sciences (internal medicine) in 1945. He served on the faculty at Columbia University College of Physicians and Surgeons, first as an instructor in medicine from 1946 to 1947, then as an associate in medicine from 1947 to 1949, an assistant professor of medicine from 1949 to 1958, and an associate professor of medicine in 1958. In 1958 Dr. Wégria was appointed professor of medicine and director of the Department of Medicine at the Saint Louis University School of Medicine. He held these positions from 1958 to 1963. From 1963 to 1979, Dr. Wégria served as professor and chair of the Department of Pharmacology at the same university. He served as chair emeritus and attending physician at the Veterans Administration Hospital in St. Louis until his death in 2000. Dr. Wégria was visiting professor at Lovanium University in Leopoldville, Belgian Congo, from January to May 1958, the Arthur E. Strauss Visiting Physician at Jewish Hospital in 1960, and chairman of the dean's committee of the Veterans Administration Hospital in St. Louis from 1961 to 1962.

During his career, Dr. Wégria was a member of numerous professional and scientific societies and was a fellow of the American Association for the Advancement of Science and the New York Academy of Science. He earned many honors throughout his career; he was named as an Honorary Foreign Member of the Argentina Society for Pharmacology and Therapeutics and received a doctor of science (honoris causa) from New York Medical College in Valhalla in 1984. The René Wégria Award in Pharmacology is given annually to a graduating medical student for excellence in pharmacology at Saint Louis University School of Medicine Dr. Wégria's research was in cardiovascular physiology and pharmacology, and he was considered a world authority on the regulation of coronary circulation, ventricular fibrillation, and cardiac glycosides. He published numerous papers with the eminent physiologist Carl Wiggers and pharmacologist Gordon Moe. He published nearly ninety papers in prestigious journals such as the *American Journal of Physiology*, the *American Heart Journal*, *Journal of Clinical Investigation*, and *Circulation Research*, among others. Dr. Wégria served faithfully in his role as chair of the Department of Pharmacology at Saint Louis University School of Medicine for sixteen years.

Medical and Graduate Education

Sophomore Year

101. Essentials of Pharmacology. Lectures and conferences, six hours. Laboratory, six hours per week, fifteenth week through twenty-fifth week. Prerequisites: AN131, BCH201, a.b., PY111.

Covers the major groups of drugs used in medicine, with emphasis on physiological and biochemical principles concerned in the action, absorption, metabolism, excretion, and toxicity of drugs.

Wégria and Staff

Senior Year

As part of the senior year elective program, research electives of six weeks duration are available in a variety of topics in the Department of Pharmacology.

Graduate Courses

200. Pharmacological Principles of Human Therapeutics (2).
201. Pharmacology of the Cardiovascular System (2).
202. Pharmacology of Smooth Muscle (2).
250. Molecular Biophysics (2). Prerequisites: College physical chemistry and consent of instructor.
251. Molecular Biophysics (2).
291. Pharmacology Journal Club (1 or no credit).
293. Pharmacology Seminar (1 or no credit).
293. Research Topics (Credit to Be Arranged).
298. Graduate Reading Course (Credit to Be Arranged).
299. Thesis Research (No credit).

Faculty Research Interests

- **Irving Coret, MD, Emory University**
  Autonomic pharmacology, theoretical pharmacology, drug-receptor interactions.

- **Alvin Gold, PhD, Saint Louis University**
  Endocrine pharmacology, cellular and molecular regulatory processes.

- **Yee S. Kim, PhD, Saint Louis University**
  Molecular pharmacology, endocrine pharmacology, vitamin D control and regulation.

- **Budh Dev Bhagat, PhD, University of London**
  Adrenergic mechanisms, action of drugs on catecholamine synthesis and release.

- **Naranjan Singh Dhalla, PhD, University of Pittsburgh**
  Autonomic pharmacology and physiology, adrenergic mechanisms, catecholamine metabolism.

- **H. Hoen Oei, PhD, Saint Louis University**
  Cardiovascular pharmacology, coronary circulation, anti-arrhythmic drugs and myocardial failure.

- **René Wégria, MD, University of Liège; MedScD, Columbia University**
  Regulation of coronary circulation, ventricular fibrillation, mechanism of action of cardiac glycosides.

- **Maysie Jane Hill Hughes, PhD, Saint Louis University**
  Actions and behavior of smooth muscle following hypoxia.

- **Sham Sunder Gandhi, DVM, MP, College of Vetenarian Medicine, Mhow, India; PhD, Saint Louis University**
  Autonomic and Cardiovascular Physiology.

- **Andrew Lonigro, MD, Saint Louis University**
  Clinical pharmacology, cardiovascular pharmacology, mechanism and treatment of hypertension.

Graduates during the Wégria Era

There was an active graduate program during the Wégria era; twelve doctors of philosophy and four masters of science were awarded.

PhD Graduates
- 1963 Sarah Hooper
- 1963 Maysie Jane Hill Hughes
- 1964 Alfred D. Goldstone
- 1965 Yee Kim
- 1965 Alvin Gold
- 1966 Jack Ging Lun Jue
- 1968 Gurbachan Dess Narme
- 1969 Sham Sunder Gandhi
- 1971 Duk Joe Park
- 1971 Ludmilla Syrotenko-Getsin
- 1973 Hong Hoen Qei
- 1977 Thomas Hamilton Hale

Master of Science Graduates
- 1964 Charles E. M. Fisher
- 1965 Alice Ham-Hsisang Yang
- 1965 Jack Ging Lun Jue
- 1979 Tiber Charles Kapjas

Faculty during the Wégria Era

(By year of appointment)

Kuang-Mei Wang, PhD (1963)

Dr. Wang earned a BS from National Southwestern University of China in 1942. He followed this with an MS from Syracuse University in 1949 and a PhD from the University of Missouri in 1951. He was appointed assistant professor in pharmacology at Saint Louis University in 1963.

Maysie Jane Hill Hughes, PhD (1964)

Dr. Hughes received a BA from Washington University in St. Louis in 1960 and a PhD in pharmacology from Saint Louis University in 1963. Her research involved actions and behavior of hypoxia in smooth muscle. She was appointed instructor in pharmacology at Saint Louis University in 1964 and promoted to assistant professor in 1966.

Naranjan Singh Dhalla, PhD (1965)

Dr. Dhalla earned a BS from Panjab University in India in 1956 and an AlC from the Institute of Chemistry in India in 1961. He followed those achievements with an MS from the University of Pennsylvania in 1963 and a PhD from the University of Pittsburgh in 1965. He was appointed assistant professor in pharmacology and a research associate in biochemistry at Saint Louis University in 1965. His research was in the field of adrenergic mechanisms.

Jack Ging Lun Jue, PhD (1966)

Dr. Jue received a BS in pharmacy from the St. Louis College of Pharmacy in 1956 and an MS from the same school in 1958. He earned an MS in pharmacology at Saint Louis University in 1965 and a PhD in 1966. He was appointed instructor in pharmacology in 1966.

Yee S. Kim, PhD (1966)

Dr. Kim, originally from Seoul, Korea, earned a BS in chemistry at Kansas State University in 1957, followed by an MS in chemistry in 1960. He earned his PhD in pharmacology at the Saint Louis University School of Medicine in 1965. He followed this with a postdoctoral fellowship in 1965 and 1966 in the Department of Biology at the State University of New York, Buffalo, School of Medicine working with Dr. Hal L. Segal. Dr. Kim accepted the position of assistant professor in pharmacology at the Saint Louis University School of Medicine in 1966 under the chairmanship of Dr. René Wégria. He was promoted to associate professor in 1971 and professor in 1977. In 1985 Dr. Kim took a sabbatical year in the Department of Bone Metabolism and Endocrinology at the Washington University School of Medicine. Dr. Kim's research was in the metabolism of macromolecules and endocrine pharmacology. He was one of the first scientists to demonstrate that vitamin D acted on membrane receptors as well as the well-known nuclear receptors. He received support from the NIH and taught both graduate and medical students. He was a loyal member of the Department of Pharmacology under the chairmanship of both Dr. Wégria and Dr. Westfall and was a valuable faculty member at Saint Louis University for thirty-six years until his retirement in 2002. Dr. Kim spoke five languages, sang opera, and was a talented chef, with specialties in Korean, Japanese, and Italian food. For many years, Dr. Kim would sing "Danny Boy" in Korean at Dr. Westfall's annual St. Patrick's Day party. Dr. Kim's wife, Young, was also a faculty member at Saint Louis University and was a Foreign student advisor for many years. From 2002 until his death in 2016, Dr. Kim was professor emeritus.

Budh Dev Bhagat, PhD (1968)

Dr. Bhagat earned an MB from Punjab University in India in 1949 and a PhD from the University of London in 1960. He was appointed associate professor in both pharmacology and physiology at the Saint Louis University School of Medicine in 1968. He was promoted to professor in both departments in 1971. Dr. Bhagat's research was in the field of adrenergic mechanisms. He studied the effects of numerous drugs on the release and metabolism of catecholamines. He authored an excellent monograph entitled *Recent Advances in Adrenergic Mechanisms*, published by Charles Thomas in 1971.

Alvin H. Gold, PhD (1968)

Dr. Gold was a native of Danton, Texas. He earned his BA in zoology at the University of Texas, Austin in 1958. He earned an MA in physiology in 1961 and a PhD in pharmacology at Saint Louis University in 1965. His postdoctoral studies were at the State University of New York, Buffalo in biology in the laboratory of Dr. H. L. Segal in 1965 and 1966. From 1966 to 1968, he was an assistant professor of biochemistry at the Bowman Gray School of Medicine at Wake Forest University. He was appointed assistant professor of pharmacology at the Saint Louis University School of Medicine in 1968, associate professor in 1973, and professor in 1978. In 1971 and 1972, Dr. Gold was a US National Academy of Sciences East European Exchange Scientist Fellow, and in 1989 he did a sabbatical in the Center for Genetics in Medicine at the Washington University School of Medicine. Dr. Gold's research was centered on the regulation of carbohydrate metabolism and was funded by both the NIH and the American Diabetes Association. Dr. Gold was an active teacher and served on numerous committees, including as the chair of the Promotions Committee, School of Medicine. He also served on the Admissions Committee and was Saint Louis University's Delegate to the United States Pharmacopeia Convention. Dr. Gold served effectively in the department under Dr. René Wégria from 1964 to 1979 and Dr. Thomas Westfall from 1979 to 2000. Dr. Gold retired in 2000 and served as professor emeritus until his death in 2016.

Sham Sunder Gandhi, DVM, PhD (1971)

Dr. Gandhi earned a DVM from the MP College of Veterinarian Medicine in Mhow, India, in 1961. He earned an MS at Iowa State University in 1965 and a PhD at Saint Louis University in 1969. He was appointed assistant professor in pharmacology at Saint Louis University in 1971.

H. Hoen Oei, PhD (1972)

Dr. Oei received his BS from the Bandung Institute of Technology in Indonesia in 1957. He went on to earn an MS in 1960 and a PhD in 1973, both from Saint Louis University. He was appointed instructor in pharmacology at Saint Louis University in 1972 and served until 1979. His research involved understanding biochemical changes such as myocardial creatine phosphate and adenosine triphosphate in acutely induced myocardial failure.

Irving A. Coret, MD

See page 30.

The ALRICK B. HERTZMAN ERA
1928–1966

Biographical Sketch of Alrick B. Hertzman, PhD

Dr. Alrick Hertzman was the fourth chair of the Department of Physiology at the Saint Louis University School of Medicine. He served in this capacity for an astonishing thirty-eight years, the longest serving chair of the Departments of Pharmacology, Physiology, or Pharmacology/Physiology at the university. Dr. Hertzman received a bachelor of arts from Gustavus Adolphus College in 1919, followed by a doctor of philosophy in physiology from the University of Wisconsin in 1924. He served as an instructor and assistant professor in physiology at the University of Michigan from 1924 to 1927 and was appointed assistant professor in the Department of Physiology at the Saint Louis University School of Medicine in 1927. He was promoted to associate professor in 1932 and professor in 1940. Upon Dr. Joseph's death in 1928, Dr. Hertzman took over administration of the department and was subsequently appointed professor and chair.

Dr. Hertzman's early research interests were in proteolysis, physical chemistry of proteins, acid-base balance, and anesthesia. In 1927 and 1928 he published twelve papers in the *American Journal of Physiology* alone concerning the nervous control of the circulation. He became a world authority on measurements of body temperature regulation. He constructed a room-sized calorimeter in which temperature and humidity could be accurately controlled and made to vary independently of each other. His research on temperature regulation was extremely valuable for flights at high altitudes and space exploration. This research also led to the development of a survival suit utilized by the US Air Force to protect pilots exposed to extremely cold temperatures. This research by Dr. Hertzman was featured in a two-page picture spread in the *St. Louis Post-Dispatch* on January 30, 1955, entitled "Exploring Mysteries of Body Temperature" and illustrated with ten pictures. The design of the suit and its operation required detailed knowledge of the various factors that influence sweating, particularly the distribution over the body surface. The research on sweating and cutaneous blood flow, sponsored by the US Air Force, provided the data influencing the final design of the suit.

Dr. Hertzman was a member of numerous professional societies, including the American Physiology Society, in which he was a distinguished member of the Circulation Group, the Society for Experimental Biology and Medicine, and the American Heart Association, among twelve or so others. He was a member of the board of directors of the St. Louis Heart Association and chairman of the research committee. He served on numerous editorial boards, including *Circulation Research*, the *American Heart Journal*, and the *American Journal of Physical Medicine*. In addition, he was a regular

FIG. 17 Dr. Alrick Hertzman

member of the Cardiovascular Study Section of the National Heart, Lung, and Blood Institute (NHLBI) at the NIH from 1946 to 1950 and a member of the training committee of the NHLBI from 1963 to 1966.

Dr. Hertzman received grants from the US Public Health Service (USPHS) of the NIH, the United States Air Force, the St. Louis Heart Association, and the Arctic Aeromedical Laboratory. He was also a popular teacher and carried a heavy teaching load, including a separate physiology course that he taught from 1928 to 1966 to medical students, dental students, and nursing students, as well as a clinical physiology course for junior medics from 1926 to 1956 and a summer physiology course for medical students from 1926 to 1938.

Teaching Activities during the Hertzman Era

During the period that Dr. Hertzman was chair of the Department of Physiology, the department had a heavy teaching load. For instance, in the 1953–54 academic year, there were two courses for medical students, one for dental students, three for nursing students, and a series of courses for graduate students.

Medical Physiology

The following descriptions were obtained from various editions of the *Bulletin of Saint Louis University: Announcements of the School of Medicine* published from 1925 to 1966.

> *PY101. 4 hrs/week of lecture and 6 hrs/week of laboratory for 20 weeks*
>
> *PY102. 3 hrs/week of lecture for 13 weeks.*

There were generally 125 students who took both courses.

> *Medical Physiology Course (101). Laboratory work: the student becomes acquainted with the use of simple forms of physiological apparatus. He is given the opportunity to work out for himself, so far as time will permit, the fundamental experiments of physiology, to develop the power of accurate observation and description, the ability to arrange results in a logical order, and the judgement to draw only warranted conclusions.*

> *In the didactic work of this course, the strong tendency in recent years toward lack of interdepartmental correlation is recognized and an attempt made to link up the work in Physiology with that in other branches, both clinical and preclinical. This is done, for example, not only by application of facts in physiology to other parts of the field but also by bringing before the class from time to time instructors from the Departments of Biochemistry, Anatomy, Internal Medicine, Neurology and Psychiatry. These instructors give brief statements that are intended to help the student in correlating the facts of physiology with those given him in other departments.*

> *In order to extend this correlation, Course 102 is offered to second semester Juniors under the joint direction of the Departments of Physiology and Medicine. Papers on assigned subjects are prepared by all members of the class from material on the original literature. These papers are read before the class and criticized and discussed by staff members. The topics are arranged so that the normal physiology (and often anatomy and embryology) of a subject is given first and immediately thereafter its pathology and clinical aspects. No attempt is made to cover a large part of the field. The course has been offered for the past six years and thus far without repetition of subject matter.*

Dental Physiology

There was a separate course for dental students.

> *PY103. 4 hrs/week of lecture for 13 weeks and 6 hrs/week of laboratory for a total of 13 weeks.*

Eighty students took the course.

Undergraduate Nursing Physiology Course

There were three separate physiology courses taught to nursing students. One was taught to students at St. John's Hospital, another to students at St. Mary's Hospital, and a final one to students at Saint Louis University Hospital. There were 150 students who took those courses. The three courses were essentially identical but were taught separately at each location. The course taught to nurses at Saint Louis University Hospital (PY54) also included a laboratory.

Graduate Courses

There were five courses specifically for graduate students which included:

- PY291 4 hours/week for one semester
- PY231 2 hours/week for one semester
- PY291 1 hour/week for both semesters and summer
- PY235 Introduction to research
- PY299 Thesis research

Physiology Laboratory

There were laboratory courses for medical, dental, and Saint Louis University Hospital nursing students. As Dr. Hertzman stated in one of his communications to Dean Colbert in 1953, "[Each laboratory] requires an enormous amount of staff supervision outside of the scheduled laboratory hours." He stated that actual laboratory hours could be expressed as such:

- PY54 4 hours/15 weeks/50 students = 3,000 student hours
- PY103 6 hours/13 weeks/80 students = 6,240 student hours
- PY101 12 hours/20 weeks/125 students = 30,000 student hours
- Total laboratory instruction = 39,240 student hours
- Average instruction load per instructor = 39,240 ÷ 6 = 6,540 student hours

Despite the enormous amount of time required, members of the Department of Physiology thought that laboratory instruction in physiology was of vital importance. The following are some excerpts from an essay by Dr. Pietro Bramante, assistant professor of physiology, entitled "Classical Physiology with Modern Instrumentation," published on May 12, 1958.

> *A laboratory course of experimental physiology is one of the most important items of the first year of study in the medical curriculum, when the student is still at the stage of "basic sciences" (anatomy, biochemistry, bacteriology, etc.). In the laboratory of physiology, for the first time the future physician learns how the living organism (both man and animal) works and how it reacts to the different stimuli and conditions of the external environment.*

> *During this laboratory course, the student has to learn various techniques and procedures and has to deal with apparatus of varying degrees of complexity. He has to become familiar with several instruments which will be his tools for the practice of medicine or research. By these means he obtains a first-hand acquaintance with examples of the facts on which the various theories of medical science are based. He learns how to correlate what he hears from his teachers in the classroom with the actual results of his own experiments and observations in the laboratory.*

The essay went on to point out that in order to accomplish the goals of the laboratory,

> *An up-to-date and efficient student laboratory of physiology should be provided with modern not too complicated but still reliable instrumentation so that the student may be put in the conduction of eliciting and recording biological events with speed and accuracy, without wasting much of his time and setting up an inefficient apparatus.*

Research during the Hertzman Era

During the Hertzman era, the Department of Physiology developed a reputation for being one of the leading laboratories in the world studying temperature regulation, and Dr. Hertzman obtained numerous grants or contracts from the US air force, the USPHS, and the Arctic Aeromedical Laboratory. The contracts from the air force especially provided a great deal of funding for faculty, graduate students, and staff in the department. Although the department carried a heavy teaching load, according to Dr. Hertzman, "Research was a major focus of the department."

Medical Horizons: An ABC National Television Series

An example of the research being conducted by Dr. Hertzman and the Department of Physiology appeared on the national TV series on achievements in research entitled *Medical Horizons*. The series was sponsored by the American Medical Association and Ciba Pharmaceutical Company in Summit, New Jersey.

10/31/55 ST. LOUIS POST-DISPATCH

St. Louis U. Medical School's 'Hot Box' on TV Network Show

Climate Chamber Research on Effect of Heat and Cold on Nervous System Demonstrated.

Research in the reaction of the human nervous system to heat and cold was demonstrated last night in a nation-wide television show originating in the physiology department at St. Louis University School of Medicine, 1402 South Grand boulevard.

During the half-hour program, which was televised by the ABC network to 39 stations, Dr. Alrick B. Hertzman, director of the physiology department, explained the climate chamber, or "hot box," where the reactions of volunteer students subjected to intense heat are recorded and studied.

The program, carried here by Station KTVI (channel 36) also dealt with development of Air Force survival suits based on studies in the climate chamber and experiments in the response of blood vessels to cold.

One of the five cameras used to televise the program took the audience into the 110-degree climate chamber where "sweat capsules" were measuring the amount of perspiration coming from a student, Joseph Carron of Ste. Genevieve.

'Patterns of Sweating.'

With the aid of a student dressed in an Air Force survival suit, Dr. Hertzman illustrated developments in insulating and ventilating the suit that came as a result of studies in the chamber which have established "patterns of sweating."

Experiments in the response of blood vessels were shown in a room where graphs recorded the reaction of blood vessels in the nose and fingertips of a student seated on a chair with his feet in a pan of water.

Dr. K. B. Coldwater, St. Louis surgeon, described how the climate chamber enabled him to learn why a patient with a blood vessel disorder in his leg was not being cured.

Tests in the chamber showed the patient's leg perspired, indicating that not all of the nerve fibers had been cut in a previous operation. Further surgery was performed and the ailment was cured.

Others on Program.

Appearing on the program, besides Dr. Hertzman and Dr. Coldwater, were Dr. Iain Ferguson, associate professor of physiology, Dr. William Geber, assistant professor of physiology, Dr. Theodore Cooper, post-doctorate fellow at the university, Francis LeClair, research assistant, and six student volunteers.

The program was the eighth in a "Medical Horizon" series on achievements in research sponsored by the American Medical Association and a pharmaceutical products firm in Summit, N.J.

After the telecast, the association and the firm jointly presented a television award to the school for "outstanding contribution to public understanding of medicine." Dr. James W. Colbert Jr., dean of the school, accepted the award in ceremonies at Hotel Chase.

FIG. 18 Article on *Medical Horizons* in the *St. Louis Post-Dispatch*, October 31, 1955

The episode featuring Saint Louis University, entitled "The Hot Box," was broadcast on October 31, 1955, from the Department of Physiology and demonstrated the reaction of the human nervous system to heat and cold. Dr. Hertzman and his research team guided ABC-TV newscaster Don Goddard through the physiology laboratories and introduced him to the climate chamber, or the so-called "Hot Box," where temperature and humidity were rigidly controlled. Here student volunteers were seen undergoing the sweating process. In his experiments, Dr. Hertzman caused student volunteers to sweat continuously for as long as thirty-two hours in order to learn new facts about sweating and its causes. These studies helped the Air Force design a new survival suit, which was demonstrated in the show, as well as helping surgeons perfect the technique of sympathectomy, an operation on the peripheral nervous system. Figure 18 shows a clipping taken from the *St. Louis Post-Dispatch* describing the show.

Following the telecast, Ciba Pharmaceutical and the American Medical Association jointly presented a television award to the School of Medicine for "an outstanding contribution to the public understanding of medicine." The award was accepted by Dean James W. Colbert Jr.

Graduate Training and Degrees

Under the guidance of Dr. Hertzman, research thrived in the department while faculty maintained a heavy teaching load. In addition, the department developed an active training program for graduate students. MS and PhD programs in physiology were established. Course requirements for the PhD in physiology involved a minimum of biochemistry, anatomy (gross, histology, neuroanatomy), pharmacology, and pathology. A course in physics was desirable but not required. PY201, PY231, and PY291, as well as introduction to research (PY235) and thesis research (PY299), were required. A total of nineteen PhDs and thirteen MSs were awarded from 1929 to 1966. Ten of the early degrees included dual or triple majors, primarily with biochemistry, but also biology, bacteriology, and chemistry. Starting in 1953, the PhD was given solely in physiology. Several students received NIH or NSF predoctoral fellowships; others received funding on air force contracts. Dr. Hertzman proposed an MD/PhD program to Dean Melvin Casberg in January 1951 but this program was not established.

PhD Graduates

- 1929 Milton Levy–biochemistry, physiology, bacteriology
- 1930 Sidney R. Thayer–biochemistry, physiology, bacteriology
- 1931 Paul L. Carroll, SJ–biology, physiology

- 1931 Clara E. Graham–biochemistry, physiology, bacteriology
- 1932 Philip A. Katzman–biochemistry, physiology, chemistry
- 1934 John M. Curtis–biochemistry, physiology, bacteriology
- 1934 Louis Levin–biochemistry, physiology, chemistry
- 1934 Nelson J. Wade–biology, biochemistry, physiology
- 1935 Rev. Elmer J. Trame–biology, physiology, philosophy
- 1938 Wilfred W. Westerfield–biochemistry, physiology, chemistry
- 1953 John Wilham Cox–physiology
- 1956 Theodore Cooper–physiology
- 1956 Darrell Lawrence Davis–physiology
- 1962 Louis Salvatore D'Agrosa–physiology
- 1962 Teresamma L. Pinakatt–physiology
- 1963 Sister Wilma Marie Haslag–physiology
- 1966 Sujay Kumar Guha–physiology
- 1966 Leon Donald Prakop–physiology
- 1966 Juan T. Quejada–physiology

Master of Science Graduates
- 1928 W. Helen Denvir
- 1940 John Bertley Dillon
- 1948 Ronald Christie Deerling
- 1948 Isabelle Helene Dougherty
- 1950 Thomas King Lammert
- 1951 Russell Charles Seckendorf
- 1951 J. Shuey
- 1956 Ronald Vincent Erken
- 1964 Gerald Don Vanlandingham
- 1965 John Edward Kallal
- 1965 Nicholas William Veith
- 1966 Robert Doyle Wiesenbaugh
- 1967 Richard Parker Zucker Jr.

Faculty during the Hertzman Era

(By year of appointment)
Twenty-eight individuals served on the faculty during the thirty-eight years that Dr. Hertzman was chair of the Department of Physiology at the Saint Louis University School of Medicine. (Note that Dr. Franke also served under Dr. Joseph's tenure as chair.)

John T. Brundage, PhD (1928)
Dr. Brundage earned a BA from Indiana University in 1925 followed by an MA and a PhD from the same school in 1926 and 1928, respectively. He served as an assistant in chemistry at Indiana University from 1925 to 1927 and an assistant in physiology at the University of Illinois in 1926. He served as an instructor in the Department of Pharmacology at the Saint Louis University School of Medicine from 1928 to 1930.

James Boswell Mitchell Jr., PhD (1930)
Dr. Mitchell earned a BS at Emory University in 1923 and a PhD at the University of Chicago in 1928. He was appointed instructor in physiology at Saint Louis University in 1930.

Fred Mettler, PhD (1933)
Dr. Mettler earned a BA from Clark University in 1929, followed by an MS in 1931 and a PhD in 1933 from Cornell University. He was an assistant in bacteriology at Clark University from 1927 to 1929 and a research associate at the New York City Health Department's Laboratory of Serology and Immunology from 1929 to 1930. He served as an assistant in anatomy at Cornell University from 1930 to 1931 and as an instructor at Cornell from 1931 to 1933. He was appointed instructor in physiology at Saint Louis University in 1933. He was appointed professor of anatomy at the University of Georgia School of Medicine in 1942 and at Columbia University in 1951. He was a guest investigator at the University of Illinois, Harvard University, and Rochester University.

Orville Walters, MD (1934)
Dr. Walters earned a BA from the University of Kansas in 1927 followed by an MA in 1932 and a PhD in 1934. He served as an instructor in biology at Central College in McPherson, Kansas, from 1927 to 1928, an instructor in chemistry at Enid High School in Oklahoma in 1928 and 1929, and an instructor in physiology at the University of Kansas from 1919 to 1934. He was appointed instructor in physiology at Saint Louis University in 1934. He earned an MD from Saint Louis University in 1939 and went on to become an associate professor of psychology at the

University of Illinois from 1956 to 1958 before finally serving as president of Central College in Kansas.

Clair Raymond Spealman, PhD (1936)

Dr. Spealman earned a BS from the University of California in 1932, followed by an MA in 1933 and a PhD from that institution in 1936. From 1934 to 1936, he was a teaching assistant in physiology at the University of California. He was appointed instructor in physiology at Saint Louis University in 1936. He later became associate professor of physiology at the Medical College of Virginia and ran the Federal Aviation Agency's Safety Projects Branch in 1948.

George Louis Maison, MD (1937)

Dr. Maison earned a BA from Northwestern University in 1930 and an MD from Northwestern University in 1935. He was an instructor in physiology at the University of Wisconsin from 1935 to 1937. He did his internship at Lutheran Deaconess Hospital in Chicago from 1934 to 1935. He was appointed instructor in physiology at Saint Louis University in 1937. He later became professor and head of the Department of Pharmacology at the Boston University School of Medicine.

Lawrence W. Roth, PhD (1940)

Dr. Roth earned a BA at Battle Creek College in Michigan in 1932, followed by an MA in 1933 and a PhD in 1940 from the University of Michigan. He was assistant in physiology at the University of Michigan from 1935 to 1939. He was a fellow in physiology at Northwestern University from 1939 to 1940. He was appointed instructor in physiology at Saint Louis University in 1940. He became the assistant head of the Department of Pharmacology at Abbott Labs in 1951 and the senior pharmacologist at Riker Labs in 1958.

Kenneth E. Jochim, PhD (1942)

Dr. Jochim earned a BS at the University of Chicago in 1939, followed by a PhD in 1941. He served as an instructor in physiology at the Siebel Institute of Biotechnology from 1941 to 1942 and as a research associate in physiology at Michael Reese Hospital in Chicago during those same years. He was appointed instructor in physiology at Saint Louis University in 1942 and promoted to assistant professor in 1945. He was appointed professor and chair of the Department of Physiology at the University of Kansas School of Medicine in 1946 and served as assistant dean at the University of Kansas from 1952 to 1957.

Walter Randall, PhD (1942)

Dr. Randall earned a BA from Taylor University in Upland, Indiana, in 1938, followed by an MS in 1940 and a PhD in 1942 from Purdue University. He was a fellow in physiology at Western Reserve University from 1942 to 1943. He was appointed instructor in physiology at Saint Louis University in 1942. He was promoted to assistant professor in 1946 and associate professor in 1949. He became professor and chair of the Department of Physiology at the Stritch School of Medicine at Loyola University in Chicago in 1954 and served until 1975. Dr. Randall was an internationally recognized expert on the nervous control of the heart and cardiodynamics, physiology of the circulation, and regulation of sweating. He was a member of numerous committees of the NIH and the American Physiological Society and was elected as the fifty-fifth president of the American Physiological Society. Dr. Randall gave the first annual D'Agrosa lectureship (see page 134). He passed away in 1997.

Douglas G. Smith, PhD (1946)

Dr. Smith received a BS in 1939 and an MA from Boston University in 1940, followed by a PhD from Ohio State University in 1945. He served as a graduate assistant in zoology at Boston University from 1940 to 1941, a graduate assistant in physiology at Ohio State University those same years, and an instructor in physiology at Ohio State University from 1942 to 1946. He was appointed senior instructor in physiology at Saint Louis University in 1946.

Helge E. Ederstrom, PhD (1947)

Dr. Ederstrom earned his BS from Beloit College in 1937, followed by an MS in 1939 and a PhD from Northwestern University in 1944. He was a teaching assistant at Northwestern University from 1937 to 1941. He was an instructor in physiology/pharmacology at the University of Missouri in 1942 and 1943 and an assistant professor at the University of Missouri from 1943 to 1947. He was appointed assistant professor in physiology at Saint Louis University in 1947 and promoted to associate professor in 1951. He became professor of physiology at the University of North Dakota School of Medicine in 1952.

Clarence N. Peiss, PhD (1950)

Dr. Peiss earned a BA in 1947, an MA in 1948, and a PhD in 1949 from Stanford University. He served as a research assistant in physiology at Stanford University from 1947 to 1948 and as a research associate in pharmacology at Stanford from 1948 to 1949. He was appointed senior instructor in physiology at Saint Louis University in 1950 and became professor of physiology at the Stritch School of Medicine at Loyola University in Chicago in 1958.

Pietro Bramante, MD (1952)

Dr. Bramante earned his MD from the University of Rome in 1944 and did an internship at the Istituto de Clinica Medica at the University of Rome in 1945. He was a resident in internal medicine from 1946 to 1951 and a research fellow in internal medicine at the Karolinska University Hospital in Stockholm in 1952. He became a research associate in physiology at Saint Louis University from 1952 to 1954. He was appointed instructor in 1954 and promoted to assistant professor in 1957 and associate professor in 1961. He later was appointed professor of physiology at the University of Illinois School of Medicine.

Alfred T. Kornfield, PhD (1953)

Dr. Kornfield earned his BA from the University of Pennsylvania in 1942 and his PhD at Ohio State University in 1953. He served as a research fellow in physiology at Ohio State University from 1950 to 1953. He was appointed assistant professor in physiology at Saint Louis University in 1953.

Iain Donn Ferguson, PhD (1954)

Dr. Ferguson earned his MB and ChB in 1948 and his PhD in 1954, all from the University of Glasgow in Scotland. He served as a demonstrator and assistant in physiology at the University of Glasgow from 1949 to 1951 and as lecturer in physiology from 1951 to 1954. He was an intern in medicine at the Glasgow Royal Infirmary from 1947 to 1948 and a senior intern in gynecology in 1948. He was a resident in medicine from 1948 to 1949 and joined the faculty in physiology at Saint Louis University in 1954.

Alfred J. Rampone, PhD (1955)

Dr. Rampone earned his BA from the University of British Columbia in 1947, followed by an MA in 1950. He earned his PhD from Northwestern University in 1954. He was an assistant in zoology at the University of British Columbia from 1949 to 1950 and an assistant in physiology at Northwestern University 1952 from 1955. He was appointed instructor in physiology at Saint Louis University in 1955.

Alfred W. Richardson, PhD (1955)

Dr. Richardson earned a BEd at Southern Illinois University in 1940 followed by an MA and a PhD from the University of Iowa in 1947 and 1949, respectively. He served as an instructor in physiology and pharmacology at the Bowman Gray School of Medicine at Wake Forest University from 1949 to 1951 and assistant professor at Indiana University School of Medicine from 1951 to 1955. He was appointed associate professor in physiology at the Saint Louis University School of Medicine in 1955 and promoted to professor in 1959. He was appointed professor and head of the Department of Physiology at Southern Illinois University in 1966. Dr. Richardson was an electronics expert and invented many scientific instruments to measure blood flow and blood clotting. He was a consultant to both McDonnell Aircraft Corporation and NASA.

William F. Geber, PhD (1956)

Dr. Geber earned a BA from Dartmouth University in 1942 followed by an MS and a PhD from Indiana University in 1950 and 1954, respectively. He was a graduate assistant at Indiana University from 1948 to 1953 and an associate in physiology from 1953 to 1954. He served as a research associate in physiology at the University of Minnesota from 1954 to 1955 and was appointed assistant professor in physiology at Saint Louis University in 1956. He became an associate professor of medicine at the University of South Dakota School of Medicine in 1958.

Barbara R. Landau, PhD (1956)

Dr. Landau earned her BS from the University of Wisconsin in 1945, followed by an MS in 1949 and a

PhD from the University of Wisconsin. She served as an instructor in physiology at Mount Holyoke College from 1949 to 1951. She was a teaching assistant at the University of Wisconsin from 1947 to 1949 and at the University of Washington from 1951 to 1954. She was appointed instructor in physiology at Saint Louis University in 1956.

Leo C. Senay Jr., PhD (1957)

Dr. Senay earned a BA from Harvard University in 1949 and a PhD in physiology at the University of Iowa in 1957. He accepted a position as instructor in the Department of Physiology at the Saint Louis University School of Medicine in 1957 and was promoted to assistant professor in 1959. He became an associate professor in 1962. He served under four chairs, including Dr. Alrick B. Hertzman, Dr. Alexander Lind, and Dr. Mary Ruh in the Department of Physiology and Dr. Westfall in the Department of Pharmacological and Physiological Science. He served as acting chair of the department from 1966 to 1968. From 1970 to 1971, Dr. Senay took a sabbatical leave at the Human Science Laboratory of the University of the Witwatersrand in Johannesburg, South Africa, working with Professor Cyril Wyndham. Dr. Senay did another sabbatical in 1978 and 1979 at the John B. Pierce Foundation Laboratory in New Haven, Connecticut, working with Dr. John T. Stitt. Dr. Senay's research was in environmental and cardiovascular physiology. He was very active as a teacher of medical, nursing, dental, allied health, and graduate students and served on many committees locally and nationally. He was also active in the American Physiology Society and was internationally recognized for his work in environmental and exercise physiology. Dr. Senay retired in 1994 and has since served as emeritus professor.

Louis D'Agrosa, BS, PhD (1963)

Dr. D'Agrosa was born in Brooklyn, New York, and earned a BS from the City College of New York. He went on to receive his PhD in physiology from the Saint Louis University School of Medicine in 1962 and was appointed an instructor there the following year. He was promoted to assistant professor in 1965. Dr. D'Agrosa was a faculty member in the Department of Physiology under the chairmanships of Dr. Hertzman and Dr. Lind until his untimely death in 1983. He was a popular teacher and mentor to medical, dental, nursing, and allied health students. Dr. D'Agrosa's research centered around factors that control smooth muscle activity in blood vessels. He devised a research technique that involved an adaptation of photoelectric photometry coupled to a microscope to study the microcirculation using bat wings. His use of bats instead of frogs for cardiovascular system research in medical student laboratories was unique to Saint Louis University. In order to secure bats for his research, Dr. D'Agrosa become an adept spelunker and was an honorary member of the Middle Mississippi Valley Grotto, which sponsored frequent cave mapping and exploring expeditions throughout Missouri. On these trips, Dr. D'Agrosa collected the bats that he needed for his research and teaching laboratories. The bats were kept in cold rooms at five degrees centigrade until needed. The medical students nicknamed Dr. D'Agrosa "Batman" and placed a picture and sign on the door to his office. Due to his popularity as a teacher and investigator, a special endowed annual lectureship, "The Louis D'Agrosa Memorial Lecture," was established by members of the department, Saint Louis University, and friends. This lectureship has become one of the most important in the Department of Physiology and later the Department of Pharmacological and Physiological Science (see page 134).

Julius Hanson, PhD (1963)

Dr. Hanson earned a BS from North Carolina State University in 1943 and a PhD from the University of California in 1959. He was appointed assistant professor of physiology at Saint Louis University in 1963.

Sister Wilma Marie Haslag, SSM, PhD (1963)

Dr. Haslag received her BS from Saint Louis University in 1957 and her PhD in 1963. She was appointed instructor in physiology at Saint Louis University in 1963. Sister Haslag carried out research on the relationship of the menstrual cycle to responses to heat stress.

Arthur Nunn Jr., PhD (1963)

Dr. Nunn received his BS from Kansas State University in 1955, followed by an MS and a PhD from the University of Iowa in 1959 and 1960, respectively. He was a fellow in physiology at the University of Illinois from

1955 to 1956 and the University of Iowa from 1958 to 1960. He was appointed instructor of physiology at the Stritch School of Medicine in Chicago in 1960 and was appointed instructor in physiology at the Saint Louis University School of Medicine in 1963.

Howard Yanof, PhD (1963)

Dr. Yanof earned a BS from Ohio State University in 1953 and then received his MS in 1956 and his PhD in 1961 from the University of California, Berkeley. He was appointed assistant professor of physiology at Saint Louis University in 1963.

Theodore (*Ted*) Cooper, MD, PhD (1965)

Dr. Cooper earned both his PhD in physiology in 1956 and his MD in 1958 from the Saint Louis University School of Medicine. He did a residency in surgery and was professor in the Departments of Physiology and Surgery from 1965 to 1966. He also served as professor and chair of the Department of Pharmacology at the University of New Mexico, associate director and head of the NHLBI, dean of the School of Medicine at the Cornell University School of Medicine, director of the Food and Drug Administration, and, finally, the CEO and President of the Upjohn Pharmaceutical Company in Michigan.

Martha Schwandt Ellert, PhD (1967)

Dr. Ellert earned her BS from Barry College in 1962 and her PhD from the University of Miami in 1967. She was appointed instructor in physiology at Saint Louis University in 1967.

Florent E. Franke, MD

See page 22.

Faculty Members that Went On to Leadership Roles

Numerous members of the faculty who served in the department during the Hertzman era went on to serve in important leadership roles in academia, government, and industry.

Name	Leadership Position
Orville Walters	President, Central College, McPherson, Kansas
Clair R. Spealman	Chief, Safety Projects Branch, Federal Aviation Agency
Walter Randall	Chair, Department of Physiology, Stritch School of Medicine, Loyola University Chicago; President, American Physiological Society
Kenneth E. Jochim	Chair and Assistant Dean, Department of Physiology, University of Kansas School of Medicine
Alfred W. Richardson	Head, Department of Physiology, Southern Illinois University School of Medicine
Leo C. Senay Jr.	Interim Chair, Department of Physiology, Saint Louis University School of Medicine
George L. Maison	Head, Department of Pharmacology, Boston University School of Medicine
Theodore Cooper	Chair, Department of Pharmacology, University of New Mexico School of Medicine; Dean, School of Medicine, Cornell University; CEO and President, Upjohn Pharmaceutical Company, Inc.

Chapter Eight
DEPARTMENT OF PHYSIOLOGY: The ALEXANDER LIND ERA
1968–1988

Biographical Sketch of Alexander Lind, DPhil, DSc

Dr. Lind was the fifth chair of the Department of Physiology at the Saint Louis University School of Medicine. He was appointed in 1968 following the two-year acting chairmanship of Dr. Leo Senay. He served in this capacity for twenty years and retired in 1988. Dr. Lind, a native of Scotland, earned a bachelor of science at University of St. Andrews in Scotland. This was followed by a doctorate in physiology in 1958 and a doctor of science in 1973, both at the University of Oxford. Prior to accepting the position of professor and chair of physiology, Dr. Lind taught a wide range of medical science undergraduate and graduate students at Oxford and Edinburgh. In addition, he was director of the Indiana University Medical Center cardiopulmonary laboratory at Wright-Patterson Air Force Base and associate professor of medicine at Indiana University. Dr. Lind's research interests revolved around the human response to environmental stress and to hot climates. While on sabbatical leave as a research fellow of the National Academy of Science at the US Army's Quartermaster Research Facility in Natick, Massachusetts, he studied thermoregulation. He was an internationally recognized authority on the mechanisms that affect cardiovascular and respiratory responses to muscular function and fatigue. Dr. Lind contributed chapters to several textbooks and coauthored a book entitled *Heat Stress and Heat Disorders* in addition to numerous papers in peer-reviewed journals. Following his appointment as chair, he continued the research in temperature regulation and exercise physiology started by Dr. Alrick Hertzman and continued to make the Department of Physiology at Saint Louis University a world leader in these fields. His research was well funded by grants from the US Air Force, the US Navy, and the NIH. He continued the strong graduate program started by Dr. Hertzman and graduated twenty-four PhD students between 1974 and 1988. He also recruited numerous faculty who distinguished themselves in their research careers. Dr. Lind served on numerous school, university, and national committees. Following his retirement from the chair in 1988, Dr. Lind and his wife, Vee Lows, moved back to Scotland. He passed away on June 3, 1990.

FIG. 19 Dr. Alexander Lind

Mission and Objectives of the Department

The mission and objectives of the Department of Physiology during the Lind era are described below:

> *Physiology has a unique position as a pivotal science which draws on anatomy, biochemistry, biophysics, pathology and pharmacology as tools to probe the function of many kinds of tissues, organs or systems. A*

complete understanding of physiology is a prerequisite for the study of the various aspects of medicine where function is abnormal. This department reflects that broad scope of interests with research at the subcellular and cellular level in endocrinology, circulatory shock and pulmonary disorders. Neurophysiological studies of muscle function and development are directed towards an understanding of muscle diseases. Other research includes muscular fatigue and the circulatory responses to environmental stimuli, such as exercise. Thus, subcellular events are integrated with whole body structure and function.

In the spring issue of *Parameters in Health Care* (Volume 4, Number 1, 1979), Dr. Lind described the following focus for the Department of Physiology he chaired:

We're here to teach students and we must do it well—good teaching in the basic sciences springs from a first-rate faculty which can also do first-rate research and maintain a graduate program of high quality. ... The department's strong dual commitment to teaching and research is reflected positively in the performance of its graduates who included physicians, physiologists, nurses and physical therapists and from the output of its broad and vigorous research program. ...
The picture here is one of a department with eager enthusiasm and abundant expertise of the highest quality with international recognition. We have a long and growing list of funded research projects. The faculty publishes, on average, one paper every 12 days in journals of high repute. I'm proud of the growth of the department over recent years and I'm delighted to be associated with this excellent faculty.

Medical and Graduate Education

Medical Physiology

This course was taken by medical students during their freshman year.

Lectures 3 hrs/week, first semester, 5 hrs/week, second semester; laboratory 12 hrs/week, second semester; conferences 2 hrs/week, second semester.

During the lecture course, students were exposed to all aspects of cellular and organ system physiology, including neurophysiology, autonomic nervous system, blood and immunity, cardiovascular

FIG. 20 Department of Physiology (ca. 1981). A. R. Lind, DSc; F. Hertelendy, PhD; L. C. Senay, PhD; T. Forrester, PhD, MD (Hons.); M. B. Laskowski, PhD; Andrew Lechner, PhD; M. F. Ruh, PhD; K. W. C. van Beaumont, PhD; C. F. Funderburk, MD; Tom Ruh, PhD; and M. S. Liu, DDS, PhD.

physiology, respiratory physiology, body fluids, renal function and acid base balance, circulatory shock, thermoregulation, exercise physiology, and gastrointestinal physiology. In the laboratory work, the students became acquainted with the use of physiological apparatus. They were given an opportunity to work out for themselves, as far as time permitted, the fundamental experiments of physiology; to develop the power of accurate observation and description, the ability to arrange results in a logical order, and the judgment to draw only warranted conclusions. Biostatistics was an integral part of this course.

The didactic work of this course directed the students' attention to the more important and well-established facts of physiology, excited interest through carefully selected applications to the problems of normal living and to those of medicine, and suggested the importance to the student of the rapidly advancing frontiers of physiological research. Correlation with other fields of knowledge was repeatedly emphasized and was aided in a practical way by the cooperation of other departments.

Graduate Program

An active graduate program existed with both master of science and PhD programs. Twenty-seven PhDs were awarded during the Lind era. Undergraduate degrees were awarded in the fields of chemistry, physics, engineering, biology, and the allied health professions. Required prerequisites in the department included principles of zoology, inorganic and organic chemistry, mathematics through differential calculus, and a minimum of one year of physics. These prerequisites had to be met by the end of the first year of full-time study or by the end of the second year of part-time study in physiology.

Master of Science (Research)

This program usually required at least two years of full-time work with a minimum of six semester hours of research. Course requirements for the MS (Research) included Physiology 506, 507, 535, and 591 and Biochemistry 501 and 503 (or equivalent). For those medical students working towards a graduate degree, the first two years of the School of Medicine curriculum fulfilled part of the credit requirements for the PhD or the MS (Research). The number of credits were determined in each case by the Graduate Student Advisory Committee and the dean of the Graduate School.

Doctor of Philosophy (PhD)

The program included both basic and elective courses in physiology as well as selected courses from other departments. Two thirds of the course program were selected from course offerings in general and comparative physiology. The selection of all courses was designed to give every student a broad base of understanding of physiology and to develop maximum achievement in individual areas of interest. Courses were selected with the approval of the student's advisor and the Graduate Student Advisory Committee. Course requirements included Physiology 506, 507, 535, 591 and Biochemistry 501 and 503 or their equivalents.

Graduate Courses

PY501 General Physiology (4)
Prerequisite: Permission of course director. Survey course for non-majors with emphasis on mammals. Fall semester.

PY502 General Physiology (2)
A continuation of PY501. Spring semester.

PY506 Graduate Physiology I (4)
Prerequisite: Permission of course director. Comprehensive course covering all major systems of normal human physiology. Designed for physiology majors.

PY507 Graduate Physiology II (7–8)
A continuation of PY506.

PY510 Statistical Methods in Biomedical Research (2)
Emphasis on experimental design and data analysis in biomedical research. Topics include probability, probability distributions, logical basis of statistical inference, and point and interval estimation. Surveys a variety of analysis of variance models, correlation and regression techniques, and nonparametric methods.

PY514 Body Fluids (4)
Seminar designed for maximum participation in considering the control, constituents and interrelationships of the various body fluids.

PY517 Control of Peripheral Blood Flow (2)
Survey of the variety of control mechanisms available for various peripheral vascular beds. Detailed study of skeletal muscle and coronary vascular beds, with special reference to applications in exercise and management of myocardial ischemia.

PY518 Neurophysiology (3)
Prerequisites include graduate level physiology and neuroanatomy, or their equivalents. Course is intended as an advanced treatment of both central and peripheral neurophysiology with an emphasis on the discussion of major original papers.

PY525 Advanced Topics in Oxygen Transport Physiology Seminar (2)
An integrative exploration of the cardiopulmonary, hematological, and cellular adaptations which promote oxygen delivery from the atmosphere to mitochondria. Presentations by faculty and students emphasize mammalian systems, but may be comparative,

developmental, environmental, or clinical in scope. Class meets for two to three hours once a week. Alternate years.

PY531 Advanced Techniques in Experimental Physiology (1–6)

PY535 Seminar in Physiology (1)

PY541 Isometric Exercise and Muscular Fatigue (4)
Advanced course involving reading, discussion, and experimental exploration of the physiological responses and their underlying mechanisms to isometric exercise and associated muscular fatigue.

PY550 Radioisotopes in Research and Medicine (2)
The theory and practical application of radioisotopes to research and medicine. Emphasis on liquid scintillation counting. Includes understanding the instrumentation, counting statistics, quenching, double label counting, the use of various types of cocktails and protein solubilizers, and Cerenkov counting. The theory of gamma counting and radiation safety.

PY552 Recent Advances in the Mechanism of Action of Steroid Hormones (2)
Lecture and student presentation course exploring the discovery and investigation of steroid receptor proteins in target issues. Emphasis on methodology and interpretation of results. Articles from recent literature selected.

PY554 Human Reproduction (2)
Comprehensive study of female and male reproductive physiology, including current problems in population growth.

PY591 Physiology Journal Club (1)

PY595 Special Study for Examinations (0)

PY597 Research Topics (1–6)
Prior permission of advisor and Graduate Student Advisory Committee required.

PY599 Thesis Research (6)

PY695 Special Study for Examinations

PPY699 Dissertation Research (12)

Graduates during the Lind Era

PhD Graduates
- 1974 Jerrold Scott Petrofsky
- 1975 John Scott Rinehart
- 1975 Ronald R. Weiss
- 1977 Carole Anne Williams
- 1977 Lawrence Joseph Baudendistel
- 1978 Stephen Gregory Hipskind
- 1979 Suzanne Marie Fortney
- 1979 Mark George Clemens
- 1980 David Miles Wood
- 1980 William Frederick Nicholson
- 1981 Diklagudai Roufa
- 1981 David Mark Olson
- 1981 Timothy W. Deakers
- 1981 Robert Gerald Brzyski
- 1982 Deborah Anne Roess
- 1983 Elene Masoud Awad
- 1983 Randall William Hammond
- 1983 Patrick Ross
- 1984 Jayshree Bhalabhai Desai
- 1984 Jeffery Keene
- 1986 James P. Dixon
- 1987 Mark Alan Shanfeld
- 1987 Faisal Ismail Mohammed
- 1988 Anna Marie Moudy
- 1989 Gary Richard Bergfeld
- 1989 Thomas Bernard Doyle
- 1990 Yue Lin

Master of Science (Research) Graduates
- 1973 James Cameron Strand
- 1974 Geraldine M. T. Minna
- 1976 Cynthia Mitchell Black
- 1980 Richard Thomas Bois
- 1982 Michael George Mooradd
- 1982 Joseph M. Paccinelli

Faculty Research Interests

During the Lind era, the department's research program concentrated on three major areas: 1) human physiology, especially as it related to environmental stresses; 2) endocrinology, with a strong emphasis on the actions of steroid hormones on animal tissues; and 3) muscle physiology, with a focus on the interrelationship of nerves and muscle. Extramural support (direct

TABLE 4 Extramural Support, Department of Physiology (1982–89)

Year	Direct Research Support
1982	$223,157
1983	$451,233
1984	$361,277
1985	$275,817
1986	$103,500
1987	$367,289
1988	$496,287
1989	$642,850

costs only) generated by the Department of Physiology (1982–89) is depicted in Table 4.

- Lawrence Baudendistel, MD, PhD, Saint Louis University
  The alteration of pulmonary fluid dynamics and gas exchange in the critically ill individual with pulmonary edema and the causes of increased permeability of the alveolar capillary membrane.
- Thomas Forrester, PhD, MD, University of Glasgow
  Mechanisms involved in the local regulation of skeletal muscle blood flow; extracellular nucleotides, release and action.
- Claude Gaebelein, PhD, Kent State University
  Exercise physiology with emphasis on the basis of vascular volume response in men and women.
- Frank Hertelendy, PhD, DSc, University of Reading
  Regulation of pituitary hormone secretion including hypothalamic releasing factors, prostaglandins, and cyclic nucleotides.
- Michael B. Laskowski, PhD, University of Oklahoma
  Electrophysiology and ultrastructure of neuromuscular transmission, muscle disease, mechanisms of synaptic development, developmental neurobiology.
- Andrew Lechner, PhD, University of California, Riverside
  Environmental and high-altitude physiology, cardiopulmonary adaptation in oxygen transport.
- Alexander R. Lind, DPhil, DSc, Oxford University
  Mechanisms that affect cardiovascular and respiratory responses to muscular function and fatigue.
- Maw-Shung Liu, DDS, Kaohsiung Medical College; PhD, University of Ottawa
  Myocardial and hepatic metabolism in endotoxic shock.
- Mary F. Ruh, PhD, Marquette University
  Steroid hormone action, glucocorticoids and B-lymphocyte differentiation, estrogen receptor heterogeneity.
- Thomas S. Ruh, PhD, Marquette University
  Subcellular mechanism of action of steroid hormones; the molecular actions of estrogens and antiestrogens at the nuclear level.
- Leo C. Senay, PhD, University of Iowa
  Environmental physiology, particularly the response of humans to exercise and heat.
- K. Willem van Beaumont, PhD, University of Indiana
  Hematological changes during exercise and training, fluid electrolyte balance and thermoregulation during space flight.
- Mary L. Ellsworth, PhD, Albany Medical College
  Regulation and function of the microcirculation; O_2 delivery and the role of the red blood cell as a regulator of microvascular perfusion.
- C. F. Funderburk, PhD, University of Tennessee, Memphis
  Exercise and respiratory physiology.
- Thomas Dahms, PhD, University of California
  Respiratory cardiovascular physiology, anesthesia, toxic effect of carbon monoxide.
- Louis D'Agrosa, PhD, Saint Louis University
  Factors controlling smooth muscle activity in blood vessels using the bat wing as a model.
- Budh Dev Bhagat, PhD, University of London
  Adrenergic mechanisms; action of drugs of catecholamine synthesis and release.

- **Edwin E. Westura, MD, Creighton University**
  Cardiovascular physiology and medicine.
- **William Hunter, PhD, Michigan State University**
  Cardiovascular and exercise physiology.
- **Sister Elizabeth Mary Burns, PhD, University of Colorado**
  Nursing physiology.
- **Roger Henry Secker-Walker, MD, University College London**
  Respiratory and pulmonary physiology.
- **Lawrence Arthur Solberg, PhD, University of California, Berkeley**
  Cardiovascular physiology.
- **Gregory Totel, PhD, University of Illinois at Urbana**
  Cardiovascular, respiratory, and exercise physiology.

Faculty during the Lind Era

(Year of appointment at Saint Louis University in parentheses.)

Mary F. (Christiano) Ruh, PhD (1971)

Dr. Mary Ruh is a native of the Chicago area. She earned a BS in medical technology in 1963, an MS in physiology in 1966, and a PhD in physiology in 1969, all from Marquette University. This was followed by postdoctoral training from 1969 to 1971 in the Department of Physiology and Biophysics at the University of Illinois at Urbana, where she also served as a research associate in the Department of Entomology. She was appointed assistant professor in the Department of Physiology at the Saint Louis University School of Medicine in 1971 under the chairmanship of Dr. Alexander Lind. She was promoted to associate professor in 1976 and professor in 1984. From 1988 to 1990, Dr. Ruh served as interim chair of the Department of Physiology after Dr. Lind's retirement. She was appointed professor in the newly formed Department of Pharmacological and Physiological Science in 1990 under the chairmanship of Dr. Thomas Westfall and served in this capacity until her retirement in 2006. Dr. Ruh took sabbatical leave in 1981 at the Mayo Clinic School of Medicine, working with Dr. D. O. Taft, and again at Texas A&M University in 1993 and 1994, working with Dr. D. Safe. Dr. Ruh's research was continuously funded by the NIH from 1974 to 2002 in the area of endocrine physiology, concentrating on the action and binding of estrogen and antiestrogen receptors. In addition, she did pioneering studies on the binding and action of the aryl hydrocarbon receptor by dioxin. In addition to her teaching of medical, graduate, and undergraduate students, Dr. Ruh served on many important committees both locally and nationally. She was chair of the First Year Curriculum Committee and the Faculty Women's Group at the Saint Louis University School of Medicine. Nationally, Dr. Ruh served on numerous NIH study sections. She was chair of the NIH Reproduction Endocrinology Study Section from 1990 to 1992, chair of the Etiology Panel, Office of Women's Health Peer Review—Breast Cancer 1995, and Chair of the Department of Defense Cancer Research Program from 1998 to 2000. Dr. Ruh received the Distinguished Alumni Award from Marquette University in 1990. She retired in 2006 and is professor emerita of the Department of Pharmacology and Physiology at the Saint Louis University School of Medicine.

Thomas S. Ruh, PhD (1975)

Dr. Thomas Ruh was a native of Spokane, Washington. He earned a BA in philosophy, biology, and education in 1963 at Gonzaga University, followed by a PhL at the same institution in 1964. He earned an MS in 1968 and a PhD in physiology in 1969 at Marquette University. This was followed by postdoctoral training from 1969 to 1971 in endocrinology at the University of Illinois at Urbana, working with Dr. Jack Gorski. He held the position of assistant professor in the Department of Obstetrics and Gynecology at the Washington University School of Medicine from 1971 to 1975 and was adjunct assistant professor in the Department of Biological Sciences at Southern Illinois University Edwardsville from 1973 to 1975. He accepted the position of assistant professor in the Department of Physiology at the Saint Louis University School of Medicine in 1975 under the chairmanship of Dr. Alexander Lind. He was promoted to associate professor in 1976 and professor in 1983. In 1980 Dr. Ruh was visiting scientist in the Department of Cell Biology at the Mayo Clinic School of Medicine. Dr.

Ruh was appointed professor in the new Department of Pharmacological and Physiological Science, which was formed in 1990 under the chairmanship of Dr. Thomas C. Westfall, and he served in this position until his retirement in 2003. Dr. Ruh received numerous grants from the NIH, which supported his research on subcellular hormone action, especially actions and function of estrogens and antiestrogens. Dr. Ruh was a loyal citizen of Saint Louis University and provided great service as a teacher, mentor, and committee member. He was course director for many years of the human physiology course for medical, graduate, and undergraduate students. Dr. Ruh retired in 2003 and served as professor emeritus until his death in 2016.

Maw-Shung Liu, DDS, PhD (1982)

Dr. Liu is a native of Taiwan and earned a DDS from Kaohsiung Medical University in Taiwan in 1964. This was followed by an MS in physiology from the University of Kentucky in 1970 and a PhD in physiology from the University of Ottawa in 1976. From 1964 to 1968, Dr. Liu was a lecturer in oral surgery at the Chinese Army Hospital and Kaohsiung Medical College Hospital in Taiwan. From 1968 to 1969, he was an intern in pathology at the University of Kentucky Medical Center. Dr. Liu held the position of instructor of physiology at the Louisiana State University School of Medicine from 1974 to 1976 and was promoted to assistant professor of physiology in 1976. He held the position of associate professor in the Department of Physiology and Pharmacology at the Bowman Gray School of Medicine at Wake Forest University from 1978 to 1982. Dr. Liu was appointed professor in the Department of Physiology at the Saint Louis University School of Medicine in 1982 under the chairmanship of Dr. Alexander Lind and became a professor in the newly formed Department of Pharmacological and Physiological Science in 1990 under the chairmanship of Dr. Thomas C. Westfall. He was a visiting professor to a number of medical schools in both Taiwan and China between 1984 and 1998. Dr. Liu's research was in the area of myocardial and hepatic metabolism in septic shock, and he received numerous grants from the NIH on both subjects. Dr. Liu is an internationally recognized expert for his work on septic shock, and he served on several NIH study sections, including the Surgery, Anesthesiology, and Trauma Study Section. In addition, he served on several editorial boards, including for the journals *Shock* and *Circulatory Shock*. Dr. Liu was also course director of several physiology courses at Saint Louis University and mentored an amazing thirty-four postdoctoral fellows and visiting scientists. In 2012 Dr. Liu returned to Taiwan, where he has continued his outstanding research career on the mechanisms and treatment of septic shock.

Karol Willem van Beaumont, PhD (1968)

Dr. van Beaumont is a native of the Netherlands. He earned a BS in The Hague, Netherlands, in 1955. During this period, he was a member of the Dutch national men's field hockey team. He came to the United States and earned a PhD in exercise physiology at Indiana University in 1965. From 1964 to 1966, he was a member of the Department of Physiology and Anatomy at Indiana University. Dr. van Beaumont was appointed assistant professor in the Department of Physiology at the Saint Louis University School of Medicine in 1968 under the chairmanship of Dr. Alexander Lind and was later promoted to associate professor. He became a member of the Department of Pharmacological and Physiological Science after its formation in 1990 under the chairmanship of Dr. Thomas C. Westfall. In the late 1970s and early 1980s Dr. van Beaumont had the distinction of being coach of Saint Louis University's women's field hockey team, which reached the rank of number one in the country in several polls and on several occasions. Dr. van Beaumont's research was in the area of environmental and exercise physiology stress and in orthostatic intolerance. He taught numerous undergraduate, nursing, and allied health students, and his laboratory exercises were very popular. Dr. van Beaumont retired in 1995 and moved to California.

Thomas Forrester, PhD, MD (Hons) (1975)

Dr. Forrester was a native of Glasgow, Scotland, and earned a ChB from the University of Glasgow in 1962, followed by a PhD in physiology in 1967. He was also awarded an MD with honors from the University of Glasgow in 1982. Dr. Forrester did postdoctoral studies in the laboratory of Professor Sir Bernard Katz in the

Department of Biophysics at University College London from 1968 to 1969. Sir Bernard was awarded a Nobel Prize in Physiology or Medicine in 1970 for his work on neurotransmission. Dr. Forrester held the position of lecturer at the Institute of Physiology of the University of Glasgow from 1963 to 1971 and senior lecturer at the same institution from 1971 to 1975. He came to the United States in 1975 to work as an associate professor in the Department of Physiology at the Saint Louis University School of Medicine under the chairmanship of Dr. Alexander Lind. He was appointed to the position of secondary assistant professor in the Department of Medicine at the Saint Louis University School of Medicine in 1983. Upon the formation of the Department of Pharmacological and Physiological Science in 1990, under the chairmanship of Dr. Thomas Westfall, Dr. Forrester was appointed associate professor and ultimately professor in that department. Dr. Forrester was a visiting professor in the Department of Medicine at the Indiana University School of Medicine in 1968 and a research consultant to the British Council in the Department of Biophysics at the University of Saarland in Homburg, West Germany, in 1971. He served as external examiner for the PhD in physiology at the University of London in 1974 and at King Abdulaziz University in Jeddah, Saudi Arabia, in 1985 and 1989. Dr. Forrester's research studied extracellular nucleotide release and actions in the cardiovascular system. He did pioneering work demonstrating that adenosine triphosphate (ATP) release from muscle had important actions as a mediator on blood flow and vascular smooth muscle contraction and relaxation. His research was supported by the Medical Research Council and the American Heart Association. Dr. Forrester was a prolific teacher and lecturer of medical, graduate, and undergraduate students and mentored many PhD students. He organized and conducted many mammalian laboratories that were well received by various student groups. One of the medical classes voted him "Best Dressed Professor of the Year," beating out Dr. Westfall, who finished in second place. Dr. Forrester retired in 2006 and served as professor emeritus until his death in 2016.

Mary L. Ellsworth, PhD (1988)
Dr. Mary Ellsworth is a native of New York. She earned a BA in biology at the University of Connecticut in 1972, an MS in biology (physiology) at the University of Hartford in 1978, and a PhD in physiology at Albany Medical College in 1981 under the mentorship of Dr. Roy D. Goldfarb. She did postdoctoral studies in the Department of Biomedical Engineering at Rensselaer Polytechnic Institute in Troy, New York, in 1981 and 1982, followed by a postdoctoral fellowship in the Department of Physiology at the Medical College of Virginia, Virginia Commonwealth University School of Medicine from 1982 to 1985. She was appointed research associate in the Department of Physiology at the Medical College of Virginia in 1985 and became an assistant professor at that institution the following year. In 1988 she was appointed assistant professor in the Department of Physiology at the Saint Louis University School of Medicine under interim chair Dr. Mary Ruh and was appointed assistant professor in the new Department of Pharmacological and Physiological Science at Saint Louis University upon its formation in 1990 under the chairmanship of Dr. Thomas C. Westfall. She was promoted to associate professor in 1992 and professor in 1998. Dr. Ellsworth's research was in the regulation and function of microcirculation, and she was an internationally recognized expert on O_2 delivery and the role of the erythrocyte as a regulator of microvascular perfusion. She was continuously funded by the NIH, including a Research Career Development Award from the NHLBI. She was very active in the Microcirculatory Society and held several leadership roles, including member of the executive council and secretary and chairman of the nominating committee, and she served on several study sections for the NIH and various foundations. She was an active teacher of medical, graduate, and undergraduate students and mentored many PhD, undergraduate, and medical students as well as serving on numerous departmental, school, and university-wide committees. Dr. Ellsworth retired in 2016 and is currently professor emerita. She lives in Colorado.

Andrew John Lechner Jr., PhD (1981)
Dr. Lechner was born in Lorain, Ohio, and earned a BS in biology (magna cum laude) from the University of Notre Dame in 1972. This was followed by a PhD (summa cum laude) in biology at the University of California, Riverside in 1977. He did postdoctoral

studies in the Department of Physiology at the University of Colorado School of Medicine from 1977 to 1981, and he accepted the position of assistant professor in the Department of Physiology at the Saint Louis University School of Medicine under the chairmanship of Alexander Lind in 1981. He was promoted to associate professor in 1985. Upon the formation of the Department of Pharmacological and Physiological Science under the chairmanship of Dr. Thomas C. Westfall, Dr. Lechner was appointed associate professor in 1990 and promoted to professor in 1994. He also holds the secondary position of professor in the Division of Pulmonary, Critical Care, and Sleep Medicine in the Department of Internal Medicine at the Saint Louis University School of Medicine. Dr. Lechner's research is in the area of pulmonary physiology, and he is an internationally recognized authority on the immunophysiology of sepsis. His research has been supported by the NHLBI, the American Lung Association, the American Heart Association, and private industry. Dr. Lechner has served both locally and nationally on numerous committees, often in a leadership capacity, directing the medical physiology course in 1996 and 1997 and the Respiration Module from 1997 to the present, serving on the USMLE Step 1 Physiology Committee of the National Board of Medical Examiners from 2000 to 2006, and serving on the National Education Committee of the American Physiological Society from 1997 to 2001. Dr. Lechner has mentored numerous PhD, medical, and undergraduate students and was the director of the MD/PhD program at the Saint Louis University School of Medicine from its inception in 1986 until 2015. He is an outstanding teacher and has received numerous awards for his teaching excellence, including two Golden Apple Awards as Best Teacher in Basic Science from graduating medical students in 2010 and 2011 and the Distinguished Teacher Award for Humanism in Basic Science from the medical class of 2008. Dr. Lechner continued his outstanding scholarly career in the Department of Pharmacology and Physiology under the direction of Dr. Thomas Burris and interim chairs Dr. Mark Voigt and Dr. Daniela Salvemini.

Michael Laskowski, BS, MS, PhD (1976)

Dr. Laskowski earned a BS from Loyola University in Chicago in 1966 and an MS and a PhD in neurophysiology from the University of Oklahoma. Dr. Laskowski did postdoctoral studies at Northwestern University and Vanderbilt University and was an assistant professor at Vanderbilt before accepting a position at Saint Louis University in physiology in 1976. He was a neurophysiologist who studied the physiology and ultrastructure of neuromuscular transmission, muscle disease mechanisms of synaptic development, and developmental neurobiology. He was course director of the medical physiology course and a popular teacher. He was well known for his colorful ties, which were often displayed by medical students in the library. He was awarded several Golden Apple awards from graduating medical students for his excellence in teaching. He also served as assistant dean for preclinical students. He left Saint Louis University to accept the position of coordinator of the WWAMI (Washington, Wyoming, Alaska, Montana, and Idaho) Regional Medical Education Program of the University of Washington.

Claude J. Gaebelein, PhD

Dr. Gaebelein earned his BA from John Carroll University in 1966 and an MS and a PhD from Kent State University in 1968 and 1971, respectively. He studied exercise physiology with an emphasis on the basis of vascular volume responses in men and women.

Thomas E. Dahms, PhD (1974)

Dr. Dahms earned a BA from the College of Wooster in 1964 and a PhD from the University of California in 1970. He was appointed assistant professor in physiology at Saint Louis University in 1974. He carried out research in the areas of respiratory and cardiovascular physiology, anesthesia, and toxic effects of carbon monoxide. He also studied the effects of passive smoking in asthmatics and mechanisms for the formation of pulmonary edema. He subsequently took an appointment as head of research in the Department of Anesthesiology at Saint Louis University.

John S. Reinhart, PhD (1976)

Dr. Reinhart earned his BA from Wesleyan University in 1970 and his MA from Miami University in 1972. He received a PhD from Saint Louis University in 1975. He was appointed instructor in physiology at Saint Louis University in 1976.

Cullie F. Funderburk, PhD (1968)
Dr. Funderburk earned a BS from Wake Forest University and a PhD from the University of Tennessee's medical school in Memphis. He was appointed assistant professor of physiology at Saint Louis University in 1968 and carried out research in exercise and respiratory physiology.

Edwin Eugene Westura, MD (1971)
Dr. Westura received an AG from St. Peter's College in Jersey City, New Jersey, in 1952, followed by an MD at the Creighton University School of Medicine in 1957. He was appointed associate professor in the Departments of Internal Medicine and Physiology at Saint Louis University in 1971. He was also head of the Section on Cardiovascular Disease in the Department of Internal Medicine.

Frank Hertelendy, PhD, DSc (1971)
Dr. Hertelendy earned a BS from the University of Budapest in 1955 and a PhD from the University of Reading in England in 1962. He was jointly appointed associate professor in the Departments of Internal Medicine and Physiology in 1971. His research was in the area of reproductive endocrinology and the regulation of pituitary hormone secretion, including hypothalamic releasing factors, prostaglandins, and cyclic nucleotides. After the formation of the Department of Pharmacological and Physiological Science in 1990, Dr. Hertelendy had a secondary appointment in this new department.

William Hunter, PhD (1971)
Dr. Hunter earned his BS and MS from the University of Oklahoma in 1962 and 1965, respectively. This was followed by a PhD from Michigan State University in 1971. He was appointed assistant professor of physiology at Saint Louis University in 1971.

Sister Elizabeth Mary Burns, PhD (1972)
Sister Burns received her BS from Saint Louis University in 1954, followed by an MS at St. Johns University in 1964 and a PhD from the University of Colorado in 1971. She was appointed assistant professor of nursing and instructor of physiology at Saint Louis University in 1972 and was director of the physiology course for nurses for several years.

Roger Henry Secker-Walker, MD, MRCP (1973)
Dr. Secker-Walker earned his BA from Clare College of Cambridge University in 1956, followed by a BChir in 1959, an MD in 1960, and an MRCP in 1963 from University College London School of Medicine. He was appointed associate professor of internal medicine and physiology at Saint Louis University in 1973. He carried out research on the mechanisms for the formation of pulmonary edema.

Lawrence Arthur Solberg Jr., PhD (1972)
Dr. Solberg earned a BA and a PhD from the University of California, Berkeley in 1966 and 1971, respectively. He was appointed instructor in physiology at Saint Louis University in 1972.

Gregory Totel, PhD (1972)
Dr. Totel earned a BA from Luther College in 1967 and an MS and a PhD from the University of Illinois at Urbana in 1969 and 1971, respectively. He was appointed instructor in physiology at Saint Louis University in 1972.

Jerrold S. Petrofsky, PhD (1974)
Dr. Petrofsky earned his PhD in the Department of Physiology at the Saint Louis University School of Medicine in 1974. He was an assistant professor in the department until 1978. After leaving Saint Louis University, he ultimately became executive director of the National Center for Rehabilitation Engineering and professor of engineering physiology and computer engineering at Wright State University. He was the Alumni Merit Awardee at Saint Louis University in 1987 for his work in equipping paralyzed patients with stimulation so they would operate tricycles with their own muscles and even walk. While in the department, he also served as the undergraduate course director for the physiology course.

Leo C. Senay, PhD
See page 50.

Louis D'Agrosa, PhD
See page 50.

Budh Dev Bhagat, PhD
See page 42.

The THOMAS C. WESTFALL ERA, PART I
1979–1990

Biographical Sketch of Thomas C. Westfall, PhD

Dr. Thomas C. Westfall was the sixth chair of the Department of Pharmacology, the ninth overall, and the first William Beaumont Professor and Chair of the Department of Pharmacological and Physiological Science (Pharmacology and Physiology) at Saint Louis University School of Medicine. He served in these roles for nearly thirty-four years. Dr. Westfall is a native of Latrobe, Pennsylvania, where he attended elementary school at Holy Family Grade School and received his secondary education at St. Vincent Prep and Latrobe High School, graduating in 1955. He attended West Virginia University in Morgantown on a wrestling scholarship and was captain of the varsity team from 1957 to 1959. He won the Southern Conference Wrestling Championship at the 123-pound class in 1957 and 1958 and qualified for the NCAA Championship three times. Dr. Westfall earned his BA in 1959, MS in 1961, and PhD in 1962, all from West Virginia University. He did postdoctoral training on a National Research Service Award from the NIH in the Department of Physiology at the Karolinska Institutet in Stockholm, Sweden. He worked in the laboratory of Professor U. S. von Euler, who won the 1970 Nobel Prize in Physiology or Medicine for his discovery and identification of noradrenaline (norepinephrine) as the neurotransmitter of postganglionic sympathetic nerves. Dr. Westfall was a faculty member for two years in the Department of Pharmacology at the West Virginia University School of Medicine before moving to the Department of Pharmacology at the University of Virginia School of Medicine where he was an assistant professor from 1965 to 1969, an associate professor from 1969 to 1974, and a professor from 1974 to 1979. During this period, he was course director of the medical pharmacology course from 1972 to 1974 and director of graduate studies from 1968 to 1979. He spent a sabbatical year as a visiting scholar in the Group of Biochemical Neuropharmacology at the Collège de France in Paris, working in the laboratory of Dr. Jacques Glowinski in 1974 and 1975.

In 1979 Dr. Westfall accepted the position of professor and chair of the Department of Pharmacology at the Saint Louis University School of Medicine. He served in this capacity for eleven years. He became the first William Beaumont Professor and Chair in the newly formed Department of Pharmacological and Physiological Science in 1990 and held this position until he stepped down from the chair in January 2013. He continued as professor until he retired in July 2016. Dr. Westfall is currently the William Beaumont Professor and Chair Emeritus.

Dr. Westfall's research interests through the years have centered around coming to an understanding of the mechanisms regulating the synthesis and release of catecholamines (dopamine, norepinephrine, and epinephrine) from the central and autonomic nervous systems; the functions and pathophysiological roles and mechanism of action of presynaptic receptors; the neuropharmacology of smoking and nicotine; the

FIG. 21 Dr. Thomas Westfall

role and mechanism of action of cotransmitters, especially neuropeptide Y (NPY), in the physiology and pathophysiology of sympathetic neurotransmission; and the role and interaction of the sympathetic nervous system, NPY, and the renin–angiotensin system in the development and maintenance of hypertension. He became an internationally recognized expert in these subjects and made numerous pioneering contributions. He was one of the first to demonstrate that nicotine and smoking tobacco produced cardiovascular effects by the release of catecholamines from the adrenal medulla and sympathetic neurons, that nicotine released dopamine and other amines from neurons in the brain, and finally that nicotine preferentially increases the neuronal firing rates from mesolimbic over nigrostriatal dopamine neurons. He characterized numerous presynaptic inhibitory and facilitory receptors modulating norepinephrine and dopamine release. His classical review on the local modulation of adrenergic neurotransmission generated over three thousand requests for reprints. He helped establish neuropeptide Y (NPY) as a cotransmitter released from sympathetic and central catecholamine neurons as well as the actions and mechanisms of action of NPY in the nervous and cardiovascular systems. He also made important contributions establishing the role of sympathetic nerve activity, NPY, and the renin–angiotensin system in experimental hypertension.

Dr. Westfall has authored or coauthored 217 publications, including 161 peer-reviewed articles, 17 invited symposium papers, 12 review articles, 25 textbook chapters (including 6 in the classic textbook *Goodman & Gilman's: The Pharmacological Basis of Therapeutics*) and edited two books and authored another. He has been awarded numerous research grants as principal investigator, including twenty-eight from the NIH, totaling more than $13 million in funding for a cumulative time of 103 years over the period from 1965 to 2015.

Dr. Westfall has served on numerous editorial boards and in a variety of service positions in academia, academic societies, and federal review panels. He has been treasurer and president of the Association of Medical School Pharmacology Chairs, chairman of the Graduate Student Convocations Committee of the American Society for Pharmacology and Experimental Therapeutics (ASPET), chairman of the Nominating Committee of ASPET, chairman of the Neuropharmacology Division of ASPET, president of the Catecholamine Club; member of the Administrative Board of the Council of Academic Societies (CAS) of the Association of American Medical Colleges (AAMC), member of the Executive Committee of the Graduate Research, Education, and Training (GREAT) Group of the AAMC, a member of the Pharmacology Test Committee of the National Board of Medical Examiners; and membership on the Experimental Cardiovascular Sciences Study Section, the Pharmacological Sciences Review Committee, and the Parent Committee of the Specialized Centers of Research in Hypertension for the National Institutes of Health (NIH). In addition, Dr. Westfall served on fifty-five special review and special emphasis Panels for the NIH. He served as an external reviewer for several academic departments and programs at institutions such as the Rutgers University/Robert Wood Johnson School of Medicine, the University of Iowa, the University of Virginia, Wayne State University, Uniformed Services University, Southern Illinois University, Creighton University, and Queens University (Canada).

Dr. Westfall has mentored twenty-seven PhD students, seventeen postdoctoral fellows, and numerous summer medical student trainees, and served on over one hundred prelim or thesis committees for PhD or MD/PhD students. He has taught forty medical school, graduate school, or service courses and was course director for medical pharmacology from 1979 to 1998 and principles of pharmacology from 1998 to 2016. He was the program director for two T32 training grants from the NIH: Training in Neuropharmacology, from the National Institute of Neurological Disorders and Stroke (NINDS), for ten years (1985–95) and Training in Pharmacological Sciences, from the National Institute of General Medical Sciences (NIGMS), for twenty-five years (1990–2015). Dr. Westfall has given fifty-seven invited presentations at national or international meetings and fifty-eight seminars at institutions in the United States, Europe, Asia, and South America, including four named lectureships. He was both a Wellcome Visiting Professor and a Grass Traveling Scientist in Neuroscience. He has received ten teaching awards from medical students at Saint Louis University including the Golden Apple Award in 1983, the MSII Teacher of the Year for 1990–91, and the Distinguished Teaching Award for Presentation Skills in Basic Science in 2011. Dr. Westfall was named Burlington Northern Scholar of the Year at Saint Louis University in 1985 and was honored by Saint Louis University by being selected as the Macebearer for the 2014 and the 2017 commencement ceremonies. He was inducted into the Academy of Educators at Saint Louis University in 2015 and the Academy of Pharmacology Educators of ASPET in 2016. Dr. Westfall is currently the William Beaumont Professor and Chair Emeritus.

Recruitment of Thomas C. Westfall

Dr. Westfall was recruited to the position of chair of the Department of Pharmacology at the Saint Louis University School of Medicine by Dean David Challoner, MD, with the help of a search committee, in 1978. Dr. Westfall received a letter informing him that he had been nominated as a candidate for the chair, after which he was contacted by Dr. Challoner. Dr. Westfall did not initially show much interest, but Dean Challoner said that he was going to be in Washington, DC, and would like to visit with Dr. Westfall in Charlottesville. Dr. Westfall felt he had nothing to lose by meeting with Dr. Challoner and agreed. The two met for breakfast at Dr. Challoner's hotel, which was less than a mile from Dr. Westfall's residence. Dr. Westfall, who was obviously interested in becoming a department chair, had previously interviewed for the chair at the University of Kentucky, Texas A&M University, the University of Arkansas, and the University of Arizona, and he was at that time being recruited by the University of California, Irvine. The meeting between Dr. Westfall and Dr. Challoner went well, increasing Dr. Westfall's interest in Saint Louis University, and he was subsequently invited to visit the school. During his stay in St. Louis, he gave a seminar on his research and talked to various faculty and chairs, including Dr. Alexander Lind of the Department of Physiology, Dr. Morton Weber of the Department of Microbiology, Dr. Robert Olson of the Department of Biochemistry, Dr. Paul Young of the Department of Anatomy, Dr. Stephen Ayers of the Department of Internal Medicine, and George Thoma, the vice president of the medical school, among others. As a result of this visit, Dr. Westfall became keenly interested in the position. Another visit was arranged on which Dr. Westfall's wife, Virginia (Gingy), was also invited. During this trip, the university offered him the position of chair of the department, and he and his wife took an initial look at housing options.

Upon returning to Charlottesville, Dr. Westfall consulted with several trusted colleagues for advice. These included Dr. Joseph Larner, his chair at the Department of Pharmacology at the University of Virginia; Dr. Norman Knorr, dean of the School of Medicine at the University of Virginia; Dr. Daniel T. Watts, Dr. Westfall's PhD mentor and at that time the dean of basic science at the Medical College of Virginia at Virginia Commonwealth University; and the two consultants to the Department of Pharmacology at Saint Louis University, Dr. J. P. Long, chair of the Department of Pharmacology at the University of Iowa School of Medicine and Dr. Norman Weiner, chair of the Department of Pharmacology at the University of Colorado School of Medicine. After much soul searching, Dr. Westfall agreed in December 1978 to accept the position. He and the university agreed that he would begin as chair of the Department of Pharmacology at the Saint Louis University School of Medicine on July 1, 1979.

The offer of the position of chair of pharmacology also consisted of the position of professor of pharmacology with tenure, a set-up package including funds for equipment, five new positions with set-up funds, some new laboratories, and renovations for both new and old laboratories.

Early Recruitments of Faculty

Between his acceptance of the position of chair and professor of pharmacology in December 1978 and his arrival on campus to begin his official work in August 1979, Dr. Westfall made many trips back to St. Louis. During these visits, numerous candidates were invited to the university as potential new faculty. This was necessary because the existing faculty was so small: only three full-time faculty members, Dr. Gold, Dr. Kim, and Dr. Coret, plus Dr. Lonigro, who held a secondary faculty position. Dr. Westfall stayed at the Cheshire Inn, which became almost a second home.

On a typical visit, he would stay in St. Louis for one week, and three candidates would visit for approximately one and a half days each. They would each give a seminar, interview with Dr. Westfall and others, and take a tour of the campus and city.

Dr. Westfall's plan was, by the time he started work as chair of the department, to have recruited three new faculty in addition to himself. This would increase the number of full-time faculty in the department to seven by the end of 1979.

Advertisements for the new positions were prepared and posted, and letters were sent to chairs of basic science departments as well as colleagues of Dr. Westfall both nationally and internationally. As soon as applications began arriving, outstanding candidates were invited to Saint Louis University for visits and interviews. The first two candidates to accept faculty positions in the Department of Pharmacology were well known to Dr. Westfall personally. Allyn Howlett was a postdoctoral fellow in the Department of Pharmacology at the University of Virginia working in Al Gilman's laboratory. Rex Wang was a postdoctoral fellow in George Aghajanian's laboratory in the Department of Pharmacology and Psychiatry at Yale University. Dr. Wang had given a seminar in the Department of Pharmacology at University of Virginia while Dr. Westfall was still on the faculty there. Both Dr. Howlett and Dr. Wang arrived in St. Louis at the same time as Dr. Westfall.

Many of the other new faculty recruits were identified initially by colleagues or response to advertisements. The third faculty recruit was Barry Chapnick, who was on the faculty of the Department of Pharmacology at Tulane University. He arrived in St. Louis in December 1979. Dr. Vincent Chiappinelli, a postdoc in Richard Zigmond's laboratory at Harvard University, was the next faculty recruit. He joined the department in 1980. He was followed by Margery Beinfeld, who was working at the NIH and arrived in 1981. In that same year, Al Poklis was given a secondary faculty appointment in the Department of Pharmacology; his primary appointment was in Pathology. In addition, Francis White received an appointment as research assistant professor. Dr. Mark Knuepfer, who was a research-track faculty member at Johns Hopkins University, was appointed assistant professor in 1986. He was followed by Rodrigo Andrade from the University of California, San Francisco in 1987. In 1986 Al Stephenson was appointed research assistant professor; David Malone was given a secondary appointment (his primary appointment was in Internal Medicine); and Thomas O'Donohue was appointed adjunct professor. Tom O'Donohue's primary affiliation was Head of CNS Research at Monsanto/Searle. In 1987 Randy Strong received a secondary appointment; his primary appointment was in the Division of Geriatrics in the Department of Internal Medicine. Dr. Thomas Lanthron also received an adjunct appointment in 1987. His primary affiliation was in the CNS Research Group at Searle.

Mission and Objectives of the Department from 1979 to 1990

The general purpose and objectives of the Department of Pharmacology as established by Dr. Westfall were to carry out state-of-the-art research, teaching, and training in pharmacological science and to provide service to other divisions of the university and to the community at large in these general areas. The philosophy of the department was to create an atmosphere and environment conducive to accomplishing these objectives. The common theme of the research efforts was to more fully understand the biochemistry, physiology, pharmacology, and pathophysiology of how cells communicate with each other, with emphasis on the nervous, cardiovascular, neuroendocrine, and endocrine systems. The objectives of the teaching program to medical students were to provide the student with an appropriate background for the practice of therapeutics (i.e., the clinical use of drugs to diagnose, prevent, treat, and cure disease) and to help the student assimilate the everchanging body of knowledge about drugs. The objective of the graduate training programs (pre- and postdoctoral) was to provide individuals the opportunity to achieve a high degree of competence in pharmacology, thus preparing them for research and teaching careers in this area of biomedical science. The service commitment to the university and community encompassed all parts of the previously described general objectives.

FIG. 22 Department of Pharmacology (ca. 1981). Standing (left to right): Al Poklis, Yee Kim, Marge Beinfeld, Rex Wang, Al Gold, Francis White. Seated (left to right): Irv Coret, Vincent Chiappinelli, Tom Westfall, Allyn Howlett.

Medical Pharmacology Course, 1979 to 1998

At the time Dr. Westfall became chair of the Department of Pharmacology, the pharmacology course for medical students needed a great deal of attention. A major focus for Dr. Westfall was to revitalize this course and make it one of the best, if not the best, basic science course in the medical school curriculum. Dr. Westfall had a great deal of experience in teaching medical students at the University of Virginia and was course director for several years. Since the course there had been well received, he was confident that he could bring about major changes at Saint Louis University. He introduced several innovations to the course, which existed from 1979 to 1998, when the curriculum was altered from a department-based curriculum to an organ-based curriculum.

The first innovation was the preparation of an in-depth syllabus that covered all lecture topics. This was prepared by the lecturers in the course and provided to the medical students free of charge (other departments prepared syllabi but charged the medical students). The syllabus chapters were updated each year. This continued until the late 1990s, when the Department of Curricular Affairs assumed the cost of the syllabus.

Other innovations included the introduction into the course of a number of new faculty recruited by Dr. Westfall, conferences at the end of each section to summarize and discuss the material, and clinical correlations where clinical faculty were invited to present patient and case studies to add relevance to the basic material presented in the course. A particularly innovative and popular addition to the course was the introduction of special topics. This first took place during the 1983–84 course. It consisted of four two-hour sessions and involved from ten to fifteen (depending on the year) faculty providing an equal number of topics. The purpose was to enable the faculty to interact with a relatively small number of students (ten to fourteen) and discuss controversial subjects relevant to an understanding of pharmacology and therapeutics. In these sessions, the students read the primary literature and made oral reports to their group. The reports often took the form of debates. These sessions allowed the students to have access to primary research articles and allowed intimate interaction with faculty that previously was handled in laboratory exercises. Another addition during the 1983–84 course was the addition of sessions during class time to discuss previous exams.

During the time from 1979 to 1998, the course was divided into the following sections:

- General Principles
- Drugs Affecting Synaptic and Neuroeffector Junctions (Autonomic and Somatic Nervous System Drugs)
- Drugs Affecting the Cardiovascular Renal System
- Drugs Affecting the Central Nervous System
- Chemotherapy of Parasitic, Microbial and Neoplastic Diseases
- Endocrine and GI Drugs

There were also lectures on toxicology. The sequence of the sections was altered from time to time. There were generally four exams plus a cumulative final, which consisted of National Board–type questions. Students were allowed to keep the exams for future study. A representative schedule and listing of topics are shown by Table 5. From 1980 to 1987, the course was presented three days a week from early September until late February. From 1987 to 1998, it was presented from mid-August until late March. The total number of contact hours was approximately 135 to 140 per course. During the period from 1979 to 1998, the medical pharmacology course was well received based on student evaluations of the course and faculty. Students also performed well on Part 1 of the exams given by the National Board of Medical Examiners.

Objectives of the Medical Pharmacology Course (1979-98)

> *Pharmacology is a study of the interactions of chemicals on the biological system. Of most relevance in the present context are those chemicals (drugs) which are useful in the diagnosis, treatment, cure and prevention of disease. These agents are among the most important and powerful tools in the physician's armamentarium and must be used wisely and with great discretion. The major problem confronting the physician is in making the proper scientific application of pharmacological principles when using therapeutic agents, that is, choosing the right drug in the right dose for one individual patient under a set of circumstances that may never be exactly duplicated.*
>
> *This course in medical pharmacology considers pharmacology to be a basic health science and like other basic sciences is intended to provide the student with the background for the clinical sciences and to help students to begin to acquire an ever-changing body of knowledge about drugs. Upon the completion of this course, the student should be familiar with all of the major classes of drugs, the rationale for their use as therapeutic agents, as well as an appreciation of the limits of their utility. More specifically the students should be able to describe and discuss the pharmacology of drugs affecting the autonomic, somatic and central nervous system, the cardiovascular, respiratory, gastrointestinal, renal/urinary, and endocrine system as well as drugs affecting parasitic, microbial and neoplastic diseases. Students should recognize the effects and actions of drugs on cells, tissue, organ systems and the whole patient. Moreover students should be able to describe and discuss the principles that affect all therapeutic agents—that is, the absorption, distribution, biotransformation and excretion, mechanism of action and adverse effects of drugs and finally the principles underlying optimal drug therapy.*
>
> *This course was designed therefore to provide the student with the background for therapeutics, that is, the clinical use of drugs to diagnose, prevent, cure and treat disease entities.*

This course was a large undertaking and required a major commitment on the part of each student. Table 5 shows a representative schedule taken from the 1988–89 class.

Graduate Program and Graduate Education

Although there were active graduate programs in other basic science departments when Dr. Westfall arrived at Saint Louis University, there was a very small graduate program in the Department of Pharmacology. It consisted of two predoctoral students and one master of science student. A major focus for Dr. Westfall was to develop a strong and vigorous graduate program in the Department of Pharmacology and later the Department of Pharmacological and Physiological Science. He was convinced that this would aid in the development of the department and greatly strengthen it. He sought and received support for the program from the administration as part of his recruitment package as chair. It was fortunate that the first four PhD students recruited all had advanced degrees or experience in the laboratory. These were Stuart Dryer, who had a master of science from the University of Arizona; Katia Gysling, who was originally from Chile and had been working at Yale University; Mark Voigt, who had a master of science from the University of Missouri; and Al Stephenson, who had a master of science from the Medical College of Wisconsin. Having these students as the first four in the department gave a kick-start to the graduate program.

TABLE 5 Representative Schedule, Medical Pharmacology Course (1988–1989)

General Principles

Date	Time	Topic	Instructor
Aug. 22	9-9:50	Introduction to Pharmacology	Dr. Westfall
	10-10:50	Drug Absorption and Distribution I	Dr. Howlett
Aug. 24	9-9:50	Drug Absorption and Distribution II	Dr. Howlett
	10-10:50	Drug Biotransformation and Excretion I	Dr. Westfall
Aug. 26	9-9:50	Drug Biotransformation and Excretion II	Dr. Westfall
	10-10:50	Mechanisms of Drug Action I	Dr. Westfall
Aug. 29	9-10:50	Pharmacokinetics and Drug Regimens	Dr. Courtney
Aug. 31	9-10:50	Clinical Stimulation Laboratory: Pharmacokinetics Applied to the Treatment of Asthma	Dr. Slavin
Sep. 2	9-9:50	Mechanisms of Drug Action II	Dr. Westfall
Sep. 7	9-10:50	Factors Modifying Drug Effects and Drug Dose	Dr. Westfall

Drugs Affecting the Endocrine System

Date	Time	Topic	Instructor
Sep. 8	9-9:50	Introduction to Endocrine Pharmacology, and Drug Dose	Dr. Beinfeld
Sep. 9	9-9:50	Parathyroid Hormone, Vitamin D and Calcitonin	Dr. Malone
Sep. 12	9-10:50	Adrenocorticoids	Dr. Beinfeld
Sep. 14	9-10:50	Thyroid Hormones & Anti-Thyroid Drugs Clinical Correlation	Dr. Howlett Dr. Silverberg
Sep. 16	9-9:50	Androgens	Dr. Gold
Sep. 19	9-10:50	Insulin & Oral Hypoglycemic Drugs	Dr. Gold
Sep. 21	9-10:50	Estrogens and Progestins	Dr. Howlett
Sep. 23	9-9:50	Conference	Drs. Howlett, Gold, Beinfeld, Malone
Sep. 26	9-12:00	EXAM 1 (RC)	

Drugs Affecting Synaptic and Neuroeffector Junctions (Autonomic and Somatic Nervous System and GI Drugs)

Date	Time	Topic	Instructor
Oct. 3	9-10:50	Drugs Affecting Synaptic and Neuroeffector Junctions I	Dr. Westfall
Oct. 5	9-10:50	Drugs Affecting Synaptic and Neuroeffector Junctions II Cholinergic Drugs I	Dr. Westfall
Oct. 7	9-9:50	Review Exam 1	Drs. Howlett, Gold, Beinfeld, Malone, Westfall, Courtney
Oct. 10	9-9:50	Autacoids I (Renin–Angiotensin)	Dr. Lonigro
	10-10:50	Autacoids II (Kinins–Prostaglandins)	Dr. Chapnick
Oct. 12	9-10:50	GI Drugs	Dr. Coret
Oct. 14	9-9:50	Conference	Drs. Westfall, Lonigro, Chapnick, Coret
Oct. 17	9-10:50	Cholinergic Drugs II	Dr. Westfall
Oct. 19	9-10:50	Adrenergic Drugs I	Dr. Westfall
Oct. 20	9-9:50	Adrenergic Drugs II	Dr. Westfall
Oct. 21	3-3:50	Conference	Dr. Westfall
Oct. 24	9-10:50	Autacoids I (Histamine–Serotonin)	Dr. Westfall
Oct. 26	9-10:50	Analgesics, Anti-Inflammatants and Anti-Pyretics	Dr. Stephenson
Oct. 27	9-9:50	Conference	Drs. Westfall, Stephenson
Oct. 28	9-9:50	Count Your Blessings	
Nov. 3	9-12:00	EXAM 2 (LRC)	

TABLE 5 Representative Schedule, Medical Pharmacology Course (1988–1989), *cont'd*

Cardiovascular and Renal Drugs: Toxicology

Nov. 7	9-10:50	Introduction to Cardiovascular Drugs Anti-Hypertensive Agents I	Dr. Lonigro
Nov. 9	9-10:50	Renal Pharmacology–Diuretics	Dr. Chapnick
Nov. 11	9-9:50	Review Exam 2	Drs. Westfall, Lonigro, Stephenson, Chapnick, Coret
Nov. 14	9-10:50	Clinical Correlation	Dr. Slavin
Nov. 16	9-10:50	Anti-Hypertensive Agents II	Dr. Lonigro
Nov. 18	9-9:50	Anti-Coagulants/Anti-Anemics	Dr. Stephenson
Nov. 21	9-10:50	Cardiac Glycosides	Dr. Lonigro
Nov. 28	9-10:50	Anti-Angina/Anti-Lipids	Dr. Lonigro
Nov. 30	9-10:50	Anti-Arrhythmics	Dr. Knuepfer
Dec. 2	9-9:50	Introduction to Special Topics	Dr. Westfall
Dec. 5	9-10:50	Toxicology I	Dr. Long
Dec. 7	9-10:50	Toxicology II	Dr. Long
Dec. 9	9-9:50	Conference	Drs. Lonigro, Knuepfer, Long
Dec. 12	9-9:50	EXAM 3 (LRC)	

Drugs Acting on the CNS

Jan. 4	9-10:50	Introduction to CNS Sedative-Hypnotics	Dr. Westfall
Jan. 6	9-9:50	Review Exam 3	Drs. Lonigro, Chapnick, Knuepfer
Jan. 9	9-10:50	General Anesthetics	Dr. Chiappinelli
Jan. 11	9-10:50	Anti-Epileptics	Dr. Chiappinelli
Jan. 13	9-9:50	Marihuana	Dr. Howlett
Jan. 17	8-8:50	Local Anesthetics	Dr. Knuepfer
Jan. 18	9-10:50	Opioids	Dr. Chiappinelli
Jan. 20	8-9:50	Special Topics I	
Jan. 23	9-10:50	Anti-Parkinson Drugs	Dr. Westfall
Jan. 25	9-10:50	Special Topics II	
Jan. 27	8-9:50	Psychopharmacology I	Dr. Andrade
Jan. 30	9-10:50	Psychopharmacology II	Dr. Andrade
Feb. 1	9-10:50	Drug Abuse	Dr. Poklis
Feb. 3	9-9:50	Conference	Drs. Andrade, Knuepfer, Westfall, Chiappinelli, Howlett
Feb. 7	1-4:00	EXAM 4 (LRC)	Drs. Westfall, Howlett, Knuepfer, Andrade, Chiappinelli

Chemotherapy of Microbial, Parasitic and Neoplastic Diseases

Feb. 20	9-10:50	Introduction to Chemotherapy, Sulfonamides	Dr. Kim
Feb. 22	9-10:50	Penicillins/Cephalosporins	Dr. Kim
Feb. 24	8-9:50	Special Topics III	
Feb. 27	9-10:50	Aminoglycosides/Tetracyclines/Chloramphenicol	Dr. Gold
Mar. 1	9-10:50	Anti-Viral /Miscellaneous Antibiotics	Dr. Coret
Mar. 3	9-9:50	Review Exam 4	Drs. Andrade, Knuepfer, Westfall, Chiappinelli
Mar. 6	9-10:50	Special Topics IV	
Mar. 8	9-10:50	Anti-Malarial/Anti-Amebic Drugs	Dr. Coret
Mar 9	8-8:50	Anti-Helminthics	Dr. Gold
Mar. 10	9-9:50	Tuberculosis /Leprosy	Dr. Kim
Mar. 13	9-10:50	Anti-Neoplastics I	Dr. Coret
Mar. 15	9-10:50	Anti-Neoplastics II	Dr. Coret

TABLE 5 Representative Schedule, Medical Pharmacology Course (1988–1989), *cont'd*

Chemotherapy of Microbial, Parasitic and Neoplastic Diseases, *cont'd*

Mar. 17	9-9:50	Conference	Drs. Coret/Kim
Mar. 20	9-10:50	Clinical Pharmacology	Dr. Lonigro
Mar. 22	9-10:50	Review	Staff
Mar. 27	1-4:00	FINAL EXAM (LRC)	

Special Topics (1988-1989)

Instructor	*Title*
Dr. Andrade	How Do Hallucinogens Work: Realities and Fantasies?
Dr. Beinfeld	New Therapeutic Uses of Stable Somatostatin Analogs
Dr. Chapnick	Factors Involved in Regulation of Vascular Reactivity
Dr. Chiappinelli	Alzheimer's Disease and CNS Cholinergic Function
Dr. Coret	Causes of Human Cancer
Dr. Druce	My Nose Is All Stuffed Up!
Dr. Howlett	Pediatric Psychopharmacology: Stimulants and the Hyperactive Child
Dr. Kim	Future of β-Lactam Antibiotics
Dr. Knuepfer	Pain: Its Origin and Control
Dr. Lonigro	Pharmacologic Intervention in Hypertensive Disease
Dr. Malone	What Is the Role of Dietary Calcium in the Prevention of Post-menopausal Osteoporosis?
Dr. Sprague	The Use of Corticosteroids in Pulmonary Disease
Dr. Stephenson	Arachidonic Acid Metabolites: Their Role in Health and Disease
Dr. Strong	Biological Factors Underlying Alcoholism
Dr. Westfall	To Salt or Not to Salt? That Is the Question

All four received their PhDs in 1985. An important goal for Dr. Westfall was to obtain a predoctoral training grant from NIH to support the program and give it prestige. He knew from past experiences that in order to be competitive for such a training grant, the department would have to demonstrate that it had a track record of training well-qualified PhD students. Dr. Westfall decided therefore to wait until this was the case. In the meantime, he put together an interdepartmental program in neuropharmacology and applied for a postdoctoral training grant from the National Institute of Neurological Disease and Stroke (NINDS), with Dr. Westfall as program director. This application seemed more reasonable to obtain since the members of the training faculty were all well funded. The training faculty consisted of Dr. Howlett, Dr. Wang, Dr. Chiappinelli, Dr. Beinfeld, and Dr. Westfall from the Department of Pharmacology; Dr. Coscia and Dr. Klein from the Department of Biochemistry; and Dr. Zahm, Dr. Tolbert, and Dr. Jacquin from the Department of Anatomy and Neurobiology. Later Dr. Knuepfer and Dr. Andrade were added to the training faculty. The application was successful and was funded for five years, from 1985 to 1990. Although the grant was modest and supported only two postdocs and one predoctoral candidate, it was a beginning. The PhD program continued to flourish, and by 1990 the department had already graduated sixteen PhDs. At this point, the department successfully renewed the neuropharmacology training grant for five more years; it now supported four postdoctoral and two predoctoral trainees.

At this time the new Department of Pharmacological and Physiological Science was formed with a new integrated graduate program consisting of the former programs in pharmacology and physiology. It was decided that it was now time to go for a predoctoral training grant in pharmacological sciences from the National Institute of General Medical Sciences (NIGMS). A training faculty of twenty-seven was assembled from the Departments of Pharmacological and Physiological Science, Biochemistry, Anatomy and Neurobiology, Pathology, Internal Medicine, and Pediatrics, with Dr. Westfall as program director. The first submission was not successful, but the application was immediately resubmitted, resulting in a site visit. This time the predoctoral training grant was successful and was awarded from 1990 to 1995. This T32 was renewed four times, through 2015, for a total of twenty-five years under the leadership of Dr. Westfall. The initial grant, from 1990 to 1995, supported four trainees per year, and subsequent grants supported six trainees per year. In any one year during this period, there were only twenty-six to twenty-eight training grants awarded in pharmacological sciences and many went to such prestigious universities as Harvard, Yale, Stanford, the University of Virginia, the University of Pennsylvania, Duke, Emory, and the University of Michigan.

The course structure for the graduate students from 1979 to 1998 consisted of medical pharmacology. Graduate students took this course, which consisted of the same course taken by medical students plus a separate weekly discussion group. After the restructuring of the medical course, a new pharmacology course (PPYG511, 512, 513, and 514) was developed for the graduate students (see pages 108–109). Graduate students also took Selected Topics in Neuropharmacology, Selected Topics in Cardiovascular Pharmacology, Selected Topics in Molecular and Biochemical Pharmacology, a Methodology Course, Journal Club, and Pharmacology Seminar. These courses were offered through the 1990 formation of the Department of Pharmacological and Physiological Science up until the restructuring of the medical school curriculum in 1998. At this time, new courses

TABLE 6 Selected Topics in Neuropharmacology, Late 1980s and Early 1990s

Tuesday 1:30–3:30   Thursday 9:00–11:00

Date	Topic	Instructor	Date	Topic	Instructor
Jan 26	General Introduction	Staff	Mar 10	Opioid Peptides	J. R. Martin
Jan 28	Structure and Function of Synapses	R. Andrade	Mar 15/17	Spring Break	
Feb 2	Electrophysiology as a Tool in Neuropharmacology	V. A. Chiappinelli	Mar 22	Peptides as Neurotransmitters II	M. M. Knuepfer
			Mar 24	Co-Transmission	R. Andrade
Feb 4	General Discussion of Transmitters	T. C. Westfall	Mar 29	Behavioral Pharmacology	J. Gibbons
Feb 9	Receptors and 2nd Messengers	R. Strong	Mar 31	Behavioral Pharmacology	J. Gibbons
Feb 11	Acetylcholine as a Transmitter	V. A. Chiappinelli	Apr 5	Neuropharmacology of Alzheimers Disease	V. A. Chiappinelli
Feb 16	Acetylcholine Receptors	V. A. Chiappinelli	Apr 7	Neuropharmacology of Affective Disorders	R. Andrade
Feb 18	5-Hydroxytryptamine as a Neurotransmitter	R. Andrade	Apr 12	Dopamine in CNS Function (Schizophrenia)	T. C. Westfall
Feb 23	Excitatory Amino Acid Neurotransmitters	T. H. Lanthorn	Apr 14	Dopamine in CNS Function (Parkinson's disease)	R. Strong
Feb 25	GABA as a Neurotransmitter	T. H. Lanthorn			
Mar 1	Regulation of Catecholamine Release	T. C. Westfall	Apr 19	Role of the Nervous System in Hypertension	M. M. Knuepfer
Mar 3	Norepinephrine and Dopamine as Neurotransmitters	T. C. Westfall	Apr 21	Student Presentations	
Mar 8	Peptides as Neurotransmitters I	M. M. Knuepfer	Apr 26	Student Presentations	

were developed for both the medical students and graduate students.

Each student pursuing a doctorate in pharmacology also had to successfully complete a written and oral qualifying examination. This requirement had to be completed by the end of the second year in the program. The student prepared two written research proposals in an abbreviated form, which were submitted to an ad hoc committee. After the committee accepted these proposals, one was selected for the student to prepare as a detailed proposal. As soon as the proposal was acceptable, the student then defended the proposed research project orally before a committee of five chaired by someone other than the PhD mentor (a more detailed description of this exercise is provided starting on page 105).

The major training in research for PhD candidates consisted of an independent research project under the supervision of a faculty advisor and a dissertation committee. This generally took two to three years of intense laboratory research. This work culminated in the preparation by the student of a written PhD dissertation. The final examination for the degree consisted of an oral defense of the dissertation by the student before his or her dissertation committee and a public seminar.

Representative Graduate Courses (1979–98)

Selected Topics in Neuropharmacology

This course consisted of lectures, discussions of the current literature, and mini-symposia. All students gave a formal presentation and wrote a paper on the topic of their choice. This course was taught in 1982, 1984, 1986, 1988, 1990, 1992, and 1994. Table 6 shows the course schedule from Spring 1988.

Selected Topics in Cardiovascular Pharmacology

This course consisted of a survey of current thinking and concepts involving the action of drugs on the cardiovascular system. Special emphasis was placed on various control mechanisms (autacoids, hormones, neurotransmitters, neuropeptides, etc.) in normal and pathophysiological conditions as well as the interaction of drugs ranging from the molecular to highly integrative systems. Students were required to participate in class discussion, present one oral report, and

TABLE 7 Selected Topics in Cardiovascular Pharmacology, Late 1980s and Early 1990s

Tuesday 1:30–3:30 p.m.			Thursday 9:00–11:00 a.m.		
Date	Topic	Instructor	Date	Topic	Instructor
Sept 6	Course Organization	Dr. A. J. Lonigro	Oct. 20	Kidney, Role in Cardiovascular Control	Dr. A. J. Lonigro
Sept 8	Introduction: Basic Circulatory Control Mechanisms	Dr. A. J. Lonigro	Oct. 27	Endothelial Cells in Vascular Smooth Muscle	Dr. B. M. Chapnick
Sept. 13	Eicosanoids I	Dr. A. Stephenson	Nov. 1	Function Inotropic Agents	Dr. A. J. Lonigro
Sept. 19	Eicosanoids II (2:00–4:00 p.m.)	Dr. A. Stephenson	Nov. 3	Calcium Channels: Agonists and Antagonists	Dr. B. M. Chapnick
Sept. 20	Renin-Angiotensin I	Dr. A. J. Lonigro			
Sept. 22	Renin-Angiotensin II	Dr. A. J. Lonigro			
Sept. 27	Biogenic Amines (5HT, histamine; kinin)	Dr. B.M. Chapnick	Nov. 8	Adenosine or ATP–Current Concepts	Dr. T. Forrester
Sept 29	Vasoactive Peptides I	Dr. M. M. Knuepfer	Nov. 10	Thrombolytic Agents	Dr. B. M. Chapnick
Oct. 4	Vasoactive Peptides II	Dr. T. C. Westfall	Nov. 22	Atrial Natriuretic Peptides	Dr. A. J. Trapani
Oct. 6	Vascular Neuroeffector Junction	Dr. T. C. Westfall			
Oct. 11	Cardiovascular Afferent Signals	Dr. M. M. Knuepfer	Nov. 29	Experimental Hypertensive Models	Dr. T. C. Westfall
Oct. 13	Central Neuronal Integration	Dr. M. M. Knuepfer	Dec. 1	Salt and Hypertension	Dr. T. C. Westfall
			Dec. 6	Student Presentations	
Oct. 18	Efferent Signals	Dr. M. M. Knuepfer	Dec. 8	Student Presentations	

write a paper. This course was taught in 1983, 1986, 1991, 1993, and 1995. Table 7 shows the course schedule from Fall 1994.

Selected Topics in Molecular and Biochemical Pharmacology

This course involved: 1) a discussion of concepts of efficacy and affinity as governing drug/hormone-receptor interaction, which included the use of common linear transforms of data and their interpretation (e.g., Schild, Hill, Scatchard, and other analyses; 2) ligand-binding studies; 3) signal transduction cascade mechanisms in metabolic regulation as illustrated by the regulation of the insulin receptor; 4) regulation of receptors associated with adenylate cyclase using the beta adrenergic receptor as a prototype; and 5) receptor–nuclear acceptor interactions such as steroid hormone receptor mechanisms using the vitamin D receptor as a prototype. Students were provided with appropriate review articles for major concepts, and active student participation consisted of presentation of assigned reading from the current scientific literature as well as writing a paper on a selected topic. This course was taught in 1985, 1987, 1989, 1992, 1994, and 1996. Table 8 shows the course schedule from Fall 1989. Table 9 shows the schedule for a methodology course (PRE677) that was taught in 1989.

Faculty Research Interests

The faculty in the Department of Pharmacology represented diversified backgrounds in biochemistry, physiology, neurobiology, medicine, and pharmacology. Active research interests encompassed autonomic, cardiovascular, endocrine, biochemical, and molecular pharmacology; neuropharmacology; psychopharmacology; toxicology; and clinical pharmacology. Studies were conducted using techniques ranging from whole animal physiology to molecular biology and biochemistry. A common theme involved the physiology, pathophysiology, and pharmacology of intracellular and intercellular communication. Research in the department integrated the advances made with simplified systems, enzymes, or receptors into more complex systems, cells, or organs, and finally into in vivo systems. This approach developed an appreciation of drug action from an effect on a receptor or enzyme to the therapeutic use of a drug in human disease states.

A summary of specific research interests of members of the department over the period from 1979 to 1990 is depicted on pages 73, 75, and 76. This is followed by Table 10, a summary of the extramural support

TABLE 8 Selected Topics in Molecular/Biochemical Pharmacology, Late 1980s and Early 1990s

Date	Topic	Instructor	Date	Topic	Instructor
Sept. 12	Principles of Receptor-Effector Action	Dr. Coret	Oct. 31	See above	”
			Nov. 3	Phosphatidyl Inositol & Ca^{2+} Transients	Dr. Kim
Sept. 15	See above	”			
Sept. 19	See above	”	Nov. 7	See above	”
Sept. 22	See above	”	Nov. 11	See above	”
Sept. 26	Ligand-Binding Studies of Receptors	Dr. Strong	Nov. 14	Cyclic Cascade Mechanisms of Regulation	Dr. Gold
Sept. 29	See above	”	Nov. 17	See above	”
Oct. 3	See above	”	Nov. 21	See above	”
Oct. 6	See above	”	Dec. 1	Receptor Desensitization/Down-regulation	Dr. Howlett
Oct. 10	See above	”			
Oct. 13	Transduction Mechanisms	Dr. Howlett	Dec. 5	See above	”
Oct. 17	See above	”	Dec. 8	See above	”
Oct. 19	See above	”	Dec. 12	Molecular Genetic Analysis of Receptors	Dr. Gold
Oct. 24	See above	”			
Oct. 26	See above	”			

TABLE 9 Methodology Course, 1989

Introduction to the Use of Animals in Research (Ethical, moral and legal obligations; sources of animals; standards of animal care)	R. Doyle	Neurochemical Techniques	T. C. Westfall
		• Catecholamine Levels and Turnover	T. C. Westfall
Physiological Techniques	B. M. Chapnick	• Neurotransmitter Uptake	T. C. Westfall
• Isolated Vascular Smooth Muscle	B. M. Chapnick	• Neurotransmitter Release in vitro	T. C. Westfall
• Perfused Arterial Vascular Bed	T. C. Westfall	• Neurotransmitter Release in vivo	T. C. Westfall, M. C. Beinfeld
• Langendorff Heart Preparation	T. C. Westfall	Electrophysiological Techniques	V. A. Chiappinelli
		• Unit Recording	M. M. Knuepfer
• Cardiovascular and Hemodynamics in the Dog	A. J. Lonigro; A. H. Stephenson	• Intracellular Recording	V. A. Chiappinelli, R. Andrade
• Freely Moving Unanesthetized Rat	M. M. Knuepfer	• Patch and Voltage Clamp	V. A. Chiappinelli, R. Andrade
Analytical Techniques	M. C. Beinfeld	Cellular Techniques	A. Howlett
• Radioisotopes and Liquid Scintillation Spectrometry	M. C. Beinfeld	• Receptor Pharmacology	R. Strong
• High Performance Liquid Chromatography	M. C. Beinfeld	• Signal Transduction Second Messengers–cAMP	A. Howlett
• Radioimmunoassay and ELISA	M. C. Beinfeld, A. H. Stephenson	• Signal Transduction Second Messengers–Ca^{2+}	Y. Kim
		• Cell Culture	A. Howlett
• Electrophoresis	A. H. Gold	• Molecular Biology	R. Strong

TABLE 10 Annual Extramural Support, Department of Pharmacology (1979–90)

Year	Extramural Support
1979	$ 86,000
1980	$ 150,000
1981	$ 300,000
1982	$ 316,507
1983	$ 639,617
1984	$ 862,058
1985	$ 856,058
1986	$ 980,999
1987	$1,295,968
1988	$1,057,857
1989	$1,283,708
1990	$2,259,257

generated by the full-time tenure-track faculty (listed as annual direct support). Indirect support is not included. Table 11 lists specific grants, funding agencies, and annual direct support generated by full-time faculty for a representative year, 1987. This is typical for the type of funding generated each year over the period from 1979 to 1990. Table 12 lists the overall accomplishments of the faculty over this period.

- **Rodrigo Andrade, PhD, Yale University**
  Neuropharmacology, electrophysiology of CNS neurons, biogenic amine nerotransmission.

- **Margery C. Beinfeld, PhD, Washington University**
  Biochemical neuropharmacology, neuropeptides, neuromodulators, cholecystokinin, motilin.

- **Barry M. Chapnick, PhD, University of Chicago**
  Cardiovascular pharmacology, pharmacology of vasoactive substances, peripheral circulation, mechanisms of arterial hypertension.

- **Vincent A. Chiappinelli, PhD, University of Connecticut**
  Neuropharmacology, cholinergic pharmacology, ganglionic neurotransmission, synaptogenesis, developmental neuropharmacology.

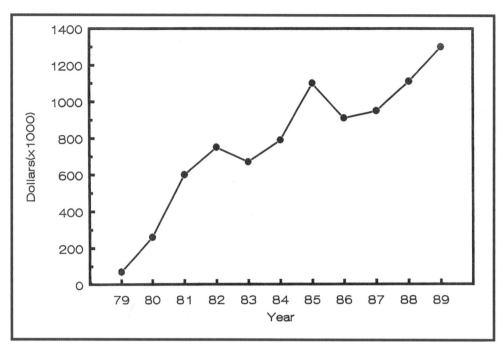

FIG. 23 Extramural support for full-time faculty (tenure track) 1979–1989.

TABLE 11 Representative Funding of Faculty as PIs for 1987

Name	Grant Number	Agency	Title	Annual Direct Costs only
Andrade, R.	R01-MH43395	ADAMHA	Mechanisms and Regulation in the CNS	$ 90,214
	AHA	American Heart Association	Serotonergic Control of Calcium Currents in Isolated Vascular Smooth Muscle Cells	$ 24,447
	PMA	Pharmaceutical Manufacturers Foundation	Identification of 5-HT Receptors in the Rat Cortex and their Mechanisms of Action	$ 10,000
Beinfeld, M.	R01-NS18667	NIH-NINCDS	Striatal CCK: Anatomy Release and Biosynthesis	$ 76,165
	1F06-TW01345	NIH Fogarty Center	Activity and Distribution of Somatrostatin Convertase	$ 13,645
Chapnick, B.	R01-HL34036	NIH NHLBI	Peripheral Vascular Control Mechanisms	$ 76,165
Chiappinelli, V.	R01-NS17574	NIH NINCDS	Neurochemistry of Autonomic Ganglia Nicotinic Receptors	$ 99,110
	R01-EY06564	NIH EY	Role of Peptides in Oculomotor Neurotransmission	$ 56,076
	INT-8518395	National Science Foundation	Purification and Characterization of Neurotoxins from the Venom of the Snake Bungarus Multicinctus Collected from the Peoples Republic of China	$ 5,417
Gold, A.	ADA	American Diabetics Association	Activator Protein of Liver Glycogen Synthase	$ 25,000

TABLE 11 Representative Funding of Faculty as PIs for 1987, *cont'd*

Name	Grant Number	Agency	Title	Annual Direct Costs only
Howlett, A.	R01-NS00868	NIH-NINCDS	Adenylate Cyclase Regulation in Neuroblastoma Cells	$ 46,480
	R01-DA03692	NIH-NIDA	Cannabinoid Actions in Neuronal Cell Brains	$ 58,524
	Unrestricted Central Research	Pfizer Inc	Study of Donated Drugs from Pfizer Central Research: Non-Classical Cannabinoid Analgesics	$ 9,800
Knuepfer, M.	R01-HL38299	NIHNHLBI	Renal Afferents Involved in Cardiovascular Regulation	$ 61,723
Lonigro, A.	SCOSR HL30572	NIHNHLBI	Adult Respiratory Failure; Project 3	$ 66,664
Malone, D.		Osteopathic Society	Expression of Chemotactic Receptors for Bone Matrix Derived Proteins on 1,25(OH)$_2$ D$_3$ Cells	$ 26,500
		VA Merit Review Grant	Developmental Biology of the Osteoclast	$ 33,754
Stephenson, A.	SCOR HL 30072	NIH-NHLBI	Adult Respiration, Failure, Project 3 Core B	$122,558
Strong, R.	R01-AC0157	NIH-NIA	Effect of Dietary Restrictions on Gene Expression	$ 74,851
		VA Merit Review	Adaption Responses to Neuronal Degeneration in the Brain During Aging	$ 50,000
Westfall, T. C.	R01-DA02668	NIH-NIDA	Acute and Chronic Nicotine on Brain Amine Release	$ 82,748
	R01-HL26319	NIH-NHLBI	Catecholamine Transmission in Vascular Smooth Muscle	$ 87,412
	R01-HL35202	NIH-NHLBI	Neuropeptide Y and Catecholamines in Hypertension	$ 73,118
	T32-NS07254	NIH-NINCDS	Graduate Training in Neuropharmacology	$ 64,652

- **Irving A. Coret, MD, Emory University**
  Autonomic pharmacology, theoretical pharmacology.

- **Alvin H. Gold, PhD, Saint Louis University**
  Endocrine pharmacology, cellular and molecular regulatory processes.

- **Allyn C. Howlett, PhD, Rutgers University**
  Biochemical neuropharmacology, cyclic nucleotide metabolism in the central nervous system.

- **Yee S. Kim, PhD, Saint Louis University**
  Molecular pharmacology, endocrine pharmacology, cellular control mechanisms.

- **Mark M. Knuepfer, PhD, University of Iowa**
  Neurophysiology, cardiovascular regulation, role of visceral afferents in autonomic control, role of CNS in hypertension, central autonomic pharmacology.

- **Andrew J. Lonigro, MD, Saint Louis University**
  Autonomic pharmacology, cardiovascular pharmacology, angiotensin and prostaglandins, hypertension.

- **David Malone, MD, Royal College of Surgeons [Ireland]**
  Nephrology, bone metabolism, developmental biology.

TABLE 12 Accomplishments of Faculty, Department of Pharmacology (1979–90)

Extramural Support Generated	$9,232,000
Peer-Reviewed Papers Published	385
Membership on Permanent NIH Study Sections	12
Participation on Special NIH Emphasis Panels	70
Participation on Non-NIH Review Panels	40
Speaking Invitations: National/International Meetings	123
Speaking Invitations: Seminars at Other Institutions	219
Membership on Editorial Boards	21
PhD Degrees Awarded	15
Participation in Medical School, Graduate School, and Service Courses	14
Course Directorships	10
Membership on National/International Committees	85
Membership on University Committees	63
Faculty Who Became Chairs of Academic Departments	4

- **Thomas L. O'Donohue, PhD, Howard University**
  Intercellular and intracellular communication, neuropharmacology, neuroendocrinology, peptides.
- **Alphonse Poklis, PhD, University of Maryland**
  Biochemical pharmacology, drug metabolism and disposition, forensic and environmental toxicology.
- **Alan H. Stephenson, PhD, Saint Louis University**
  Pulmonary pharmacology, cardiovascular pharmacology, renal pharmacology.
- **Rex Wang, PhD, University of Delaware**
  Neurotransmitters, biogenic amines, neuropharmacology, biological psychiatry.
- **Thomas C. Westfall, PhD, West Virginia University**
  Autonomic pharmacology, biochemical neuropharmacology, neurotransmitter physiology and pharmacology, monoaminergic synthesis and release.

Faculty during the Westfall Era, Part I

(Years of service at Saint Louis University in parentheses.)

Allyn Howlett, PhD (1979–2000)

Dr. Howlett is a native of upstate New York. She earned a BS from Pennsylvania State University in biochemistry in 1971, followed by a PhD in pharmacology and toxicology in 1976 from Rutgers University School of Medicine (later named the Robert Wood Johnson School of Medicine, State University of New Jersey). Her PhD mentor was Bruce Breckenridge. Dr. Howlett did postdoctoral studies in the Department of Pharmacology at the University of Virginia in the laboratory of Nobel Prize laureate Al Gilman. Gilman shared the Nobel Prize for Physiology and/or Medicine in 1994 with Martin Rodbell for the discovery of G proteins. Dr. Howlett was the first faculty member recruited by Dr. Westfall; she joined the Department of Pharmacology at Saint Louis University in 1979 as assistant professor. She was promoted to associate professor in 1985 and full professor in 1989. Dr. Howlett was an energetic teacher and mentor of PhD students. Her research was well funded by the NIH and was mainly in biochemical neuropharmacology and signal transduction. She was a pioneer in the field of cannabinoid research and made the pivotal discovery of identifying the signal transduction pathway for tetrahydrocannabinol and cannabinol drugs, that being inhibition of adenylyl-cyclase and cyclic AMP function. This set the stage for the discovery of cannabinol receptors and greatly advanced the research in this very important field. Dr. Howlett served on numerous departmental and university committees as well as numerous study sections and national committees. She left Saint Louis University in 2000 to assume the directorship of a minority training program funded by the National Institute on Drug Abuse at North Carolina Central University. She is currently a professor in the Department of Physiology and Pharmacology at the Wake Forest University School of Medicine, where she has continued her outstanding career in cannabinol research and neuropharmacology.

Rex Wang, PhD (1979–1985)

Rex Wang is a native of Taipei, Taiwan. He earned a BS in psychology from National Taiwan University in 1969 and a PhD in physiological psychology from the University of Delaware in 1975. He completed a postdoctoral fellowship in neuropharmacology at Yale University, working in the laboratory of Dr. George Aghajanian from 1974 to 1977. He was a faculty member in the Department of Psychiatry at Yale University from 1977 to 1979 before joining the Department of Pharmacology at the Saint Louis University School of Medicine in 1979. Dr. Wang's research involved studying the long-term effects of antipsychotic and psychotomimetic drugs on central catecholamine systems, particularly dopamine, using electrophysiological techniques, the interactions of cholecystokinin and dopamine, and the role of dopaminergic and cholinergic neurons in dementia, Parkinson's disease, and schizophrenia. He and his colleague, Francis White, made the important discovery of the differential actions and mechanisms of antipsychotic drugs on A_9 and A_{10} dopamine systems. He left the department in 1985 to take a position in the Department of Neuroscience at the State University of New York at Stony Brook, where he continued a successful career in neuropsychopharmacology. He is now retired.

Vincent Chiappinelli, PhD (1980–1996)

Dr. Chiappinelli is a native of New England. He earned his BA in chemistry from Boston University in 1973, followed by a PhD at the University of Connecticut in neuropharmacology in 1977, working in the laboratory of Dr. Ezio Giacobini. Dr. Chiappinelli did postdoctoral work in the Department of Pharmacology at the Harvard University School of Medicine from 1977 to 1980 in the laboratory of Dr. Richard E. Zigmond. He joined the Department of Pharmacology at the Saint Louis University School of Medicine in 1980 as an assistant professor. He was promoted to the rank of associate professor in 1985 and full professor in 1990. Dr. Chiappinelli was an outstanding teacher, mentor, and investigator. He provided valuable service to the department including serving as director of the graduate program for sixteen years. He did a sabbatical year from 1986 to 1987 in the Department of Zoology and Neurobiology at the University of Cambridge, working with Professor Eric Barnard and Dr. David Sattell. Dr. Chiappinelli was an expert in the pharmacology and physiology of autonomic ganglionic transmission and acetylcholine-nicotinic neuronal receptors. His lab purified neurotoxins termed kappa-neurotoxin from snake venoms and did pioneering studies on neuronal-nicotinic receptors using electrophysiological techniques coupled with biochemical studies. His research was well funded by the NIH, and he served on several NIH study sections and gave numerous invited lectures at national and international meetings. Dr. Chiappinelli left the department in 1996 to become chairman of the Department of Pharmacology at George Washington University School of Medicine. There, he continues his outstanding research and is currently chair of the Department of Pharmacology and Physiology and dean of research.

Barry Chapnick, PhD (1979–2006)

Dr. Chapnick is a native of Chicago. He earned a BS in pharmacy from the University of Illinois College of Pharmacy in 1965. He followed this with a PhD in pharmacology at the University of Chicago in 1971 and a postdoctoral fellowship in the Department of Pharmacology at the University of Iowa College of Medicine under the tutelage of Dr. L. S. van Orden III from 1970 to 1973. Dr. Chapnick was a faculty member in the Department of Pharmacology at the Tulane University School of Medicine from 1973 to 1979. He joined the department in 1979, the third faculty member recruited by Dr. Westfall, and served as a loyal member of the faculty until his retirement in 2006. Dr. Chapnick was a cardiovascular pharmacologist and was interested in circulatory control mechanisms, particularly the role of products of arachidonic acid, including leukotriene C_4 and D_4. Dr. Chapnick was funded through most of his career by both the NIH and the American Heart Association. After retiring in 2006, he returned to the Chicago area.

Margery Beinfeld, PhD (1981–1995)

Dr. Beinfeld is a native of Washington, DC. She earned a BS in botany from Washington University in 1968 and a PhD from that institution in biology in 1973.

She was a postdoctoral fellow in the Department of Biochemistry at the Saint Louis University School of Medicine from 1973 to 1975 in Dr. Anthony Montanosi's laboratory. She was a research associate in biochemistry and psychiatry at the Washington University School of Medicine from 1975 to 1979 and a guest worker and staff fellow at the National Institute of Mental Health (NIMH) from 1979 to 1981. Dr. Beinfeld joined the Department of Pharmacology at the Saint Louis University School of Medicine as Assistant Professor in 1981. Dr. Beinfeld's research interest was in the field of neuropeptide regulation, especially the biosynthesis, processing, and regulation of cholecystokinin. She became a nationally and internationally recognized expert in the biochemistry and molecular biology of cholecystokinin and its function and regulation. Dr. Beinfeld received strong funding from the NIH and various foundations such as the Parkinson's Disease Foundation. In 1995 Dr. Beinfeld joined the faculty of the Department of Pharmacology at the Tufts University School of Medicine, where she has continued a successful career.

Mark Knuepfer, PhD (1986–present)

Dr. Mark Knuepfer is a native of the Chicago suburbs. He earned a BA in chemistry at Grinnell College in 1975, followed by a PhD in pharmacology at the University of Iowa College of Medicine under the tutelage of Dr. Mike Brody. Mark did postdoctdoral studies at the Physiology Institute of the University of Heidelberg in Germany in the laboratory of Günter Stock from 1981 to 1982; the Department of Physiology at the University of Western Ontario in the laboratory of Dr. Francis Caleresu from 1982 and 1983; and the Department of Biomedical Engineering at the Johns Hopkins School of Medicine with Dr. Larry Schramm from 1983 to 1984. He worked as a research associate at Johns Hopkins University

FIG. 24 Department of Pharmacology, 1989

from 1984 to 1986, when he was recruited as an assistant professor in the Department of Pharmacology at the Saint Louis University School of Medicine. He was promoted to associate professor in 1989 and professor in 1996. Dr. Knuepfer has had a vigorous career as an investigator and is an expert in the central nervous system regulation of the circulation, the cardiovascular effects of cocaine, and the role of the nervous system in hypertension. In addition to his outstanding research generously funded by the NIH, Dr. Knuepfer has been an important teacher and has served in several important leadership roles at the university, including chair of the Animal Care Committee and president of the University Senate. He also served for many years as course director of the Cardiovascular Module. Dr. Knuepfer has been a visiting scientist in the Department of Physiology and Neuroscience at Beijing Medical University and a visiting scientist at the National Cardiovascular Center Research Institute in Osaka, Japan.

Dr. Knuepfer is still an active investigator, teacher, and mentor in the Department of Pharmacology and Physiology under the chairmanship of Daniela Salvemini. He is the sole remaining faculty member recruited by Dr. Westfall to the Department of Pharmacology prior to the formation of the Department of Pharmacological and Physiological Science in 1990.

Rodrigo Andrade, PhD (1987–1995)
Dr. Andrade is a native of Chile. He received a BA in biology at Kalamazoo College in Michigan in 1977. This was followed by a PhD in pharmacology at the Yale University School of Medicine in 1984, working with Benjamin Bunney and George Aghajanian. He did postdoctoral work in the Department of Pharmacology and Physiology at the University of California, San Francisco, working with Dr. Roger Nicoll from 1985 to 1987. Dr. Andrade joined the Department of Pharmacology at the Saint Louis University School of Medicine as an assistant professor in 1987 and was promoted to associate professor in 1991. Dr. Andrade's research interests were in neuro- and psychopharmacology using electrophysiological approaches. He became an expert in the regulation of neuronal excitability in various brain regions such as the hippocampus and cerebral cortex, and is a national and international authority on serotonin (5-hydroxytryptamine) receptors. He received excellent funding from both the NIH and various foundations. In 1995 Dr. Andrade joined the Department of Psychiatry and Pharmacology at the Wayne State University School of Medicine as professor and has continued his outstanding career there.

Irving Coret, MD
See page 30.

Alvin Gold, PhD
See page 42.

Yee Kim, PhD
See page 41.

Chapter Ten
The DEPARTMENT of PHARMACOLOGICAL and PHYSIOLOGICAL SCIENCE: The THOMAS C. WESTFALL ERA, PART II
1990–2013

Following the retirement of Alexander Lind, an interim chair was named for the Department of Physiology and a search committee was appointed to find a successor. Dr. Mary Ruh assumed this position and performed the role until the formation of the Department of Pharmacological and Physiological Science in October 1990. Dr. Ruh decided not to be a candidate for the permanent chair; she felt it would be a conflict of interest since her husband was a faculty member in the department. A national search was conducted, and several attractive candidates, including Dr. Vernon Bishop from University of Texas Health Science Center, San Antonio visited St. Louis. The two leading candidates withdrew their names, and after a couple of years, the search ultimately failed.

At this point, Dean William Stoneman approached Dr. Westfall about the possibility of chairing a new department resulting from a merger of the Departments of Pharmacology and Physiology. Dr. Westfall was originally leery of such an arrangement, and it was not initially well received by faculty in either department. Dr. Stoneman, who was a most persuasive person, made it difficult for Dr. Westfall not to consider the possibility of such a merger. He had available a very attractive recruitment package: six new faculty positions, monies for renovation of existing laboratories as well as creation of new labs, an endowed professorship for the new chairman, funds to expand the administrative staff and renovate the office, and funds to support the new department in terms of new equipment together with administrative support. At this juncture, Dr. Westfall consulted with colleagues nationally and talked individually with members of both departments. Several advantages of such a merger emerged: the research strengths in both departments were similar which would allow for synergism (in the Department of Pharmacology these were neuroscience, cardiovascular science, endocrinology, and signal transduction; in the Department of Physiology they were cardiovascular, respiratory, and endocrine physiology), and the requirements for both graduate programs were amazingly similar and would allow for an easy transition and synergism to strengthen both. The new department resources would allow for the expansion of the graduate programs by increasing the critical mass of students, and expansion of the faculty would allow for the recruitment of scientists trained in molecular biological approaches, which would benefit and modernize the department. Dr. Westfall was well respected as both a pharmacologist and a physiologist and had research expertise in both areas.

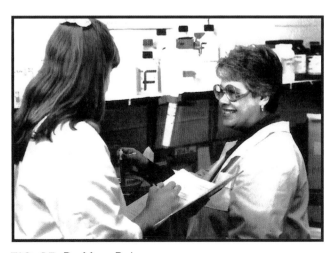

FIG. 25 Dr. Mary Ruh

St. Louis University School of Medicine is pleased to announce the formation of the department of pharmacological and physiological science. The new expanded department will carry out research, teaching and service activities formerly performed by the departments of pharmacology and physiology.

Thomas C. Westfall, Ph.D., has been appointed chairman of the new department. He has served as professor and chairman of the department of pharmacology since 1979.

In addition, Dr. Westfall now holds the newly endowed William Beaumont Chair, named in honor of the famous frontier physician and pioneer physiologist. Dr. Beaumont served as chairman of surgery in the first St. Louis University department of medicine, formed in 1837.

Having received his Ph.D. at West Virginia University, Dr. Westfall did postdoctoral and sabbatical training at Karolinska Institute in Stockholm and the College de France in Paris. He served on the faculty at West Virginia University and the University of Virginia prior to joining the faculty of St. Louis University School of Medicine.

Dr. Westfall has published over 275 research articles, reviews and abstracts. His research concerns an understanding of the mechanisms regulating the synthesis and release of catecholamines from the central and peripheral nervous system; the function and mechanisms of action of presynaptic receptors; neuropharmacology of nicotine; role of the nervous system in experimental hypertension; and the interaction of neuropeptides with catecholamine systems.

He is currently the principal investigator of several research and training grants from the National Institutes of Health and the Alcohol, Drug Abuse and Mental Health Administration.

He has served on numerous national committees including the Experimental Cardiovascular Study Section at the National Institutes of Health and the Parent Committee for the National Heart, Lung, Blood Institute's Specialized Centers of Research in Hypertension. He is currently a member of the Pharmacology Test Committee for the National Board of Medical Examiners and Chairman of the Subcommittee for Graduate Student Convocations of the American Society for Pharmacology and Experimental Therapeutics. He serves of the editorial boards of *The American Journal of Physiology* and *Blood Vessels*.

Dr. Westfall has received numerous teaching awards by medical students at St. Louis University including the Golden Apple Award. He was named Burlington Northern Scholar of the Year at St. Louis University in 1985.

The School of Medicine is pleased to have a highly respected professional lead this new era in academic and scientific growth at St. Louis University Medical Center.

William Stoneman III, M.D.
Dean
St. Louis University School of Medicine

FIG. 26 Letter from Dr. Stoneman announcing the formation of the Department of Pharmacological and Physiological Science in October 1990

The Department of Physiology, which was floundering because of the lack of a permanent chair, could move forward.

As a result of these discussions with faculty in both departments, it appeared that Dr. Westfall was acceptable as chair to both, and the faculty were willing to support such a merger. This was very important and no doubt helped with the success of the merger. Dr. Westfall reassured the faculty that this would be a new department where all would be considered equal and no one would be a second-class citizen. In addition, the disciplines of both pharmacology and physiology would be protected in the teaching programs. The decision was made that the merger would move forward with Dr. Thomas C. Westfall as the new chair, and the Department of Pharmacological and Physiological Science was officially born in October 1990.

Faculty Who Carried Over to the New Department

In addition to Dr. Westfall, members of the Department of Pharmacology who became members of the newly formed department were Al Gold, Yee Kim, Allyn Howlett, Vincent Chiappinelli, Margery Beinfeld, Barry Chapnick, Mark Knuepfer, and Rodrigo Andrade. Members of the Department of Physiology who became members of the Department of Pharmacological and Physiological Science included Charlie Senay, Mary Ruh, Tom Ruh, Willem van Beaumont, Maw-Shung Liu, Andrew Lechner, Tom Forrester, and Mary Ellsworth.

Recruitments to the New Department and Additions from Anatomy/Neuroscience

Soon after the creation of the new department, a search committee was appointed and a national search began for new faculty. Advertisements were made in key journals, and letters requesting nominations at either the assistant or associate professor positions were sent to chairs of basic science departments nationally. Recruitments were carried out for the next several years. Numerous candidates were identified, and leading candidates were invited to St. Louis for interviews. The following appointments were made:

- Nicola Partridge in 1991
- Terrance Egan in 1992
- Wendi Neckameyer in 1992
- John Chrivia in 1993
- Paul Macdonald in 1993
- Mark Voigt in 1994
- Joe Baldassare in 1996
- Medha Gautam in 1997
- Rick Samson in 1999
- Randy Sprague in 2000
- Andy Lonigro in 2000
- Karen Gregerson in 2001
- Amy Harkins in 2002
- Christian Lemon in 2008
- Decha Enkvetchakul in 2008
- Daniela Salvemini in 2009

In 2004 Dean Montelone created the Center for Anatomical Education from the previous Department of Anatomy and Neurobiology. Three of the faculty of the former department who did research in neuroscience were transferred to the Department of Pharmacological and Physiological Science. These were Dr. Scott Zahm, Dr. Mickey Ariel, and Dr. Michael Panneton.

Mission and Objectives of the Department

The Department of Pharmacological and Physiological Science was set up to have a three-fold mission: to expand the frontiers of knowledge by fostering and carrying out state-of-the-art research; to educate undergraduate, graduate, medical, nursing, and allied health students and to provide service to other divisions of the university and to the community at large. The common theme of the research efforts was to more fully understand the biochemistry, physiology, pharmacology, and pathophysiology of cellular communications, with special emphasis on the cardiovascular, endocrine, nervous, and neuroendocrine systems, and developmental biology and cancer. The objective of the teaching programs to medical students was firstly to provide the student with an appropriate base of understanding of the functions of the body, going from the cellular and subcellular levels through organs and tissues to integrated systems (by comparing normal and abnormal functions, one is better able to understand the cause of diseases and to develop therapeutic approaches to their treatment); and secondly, to provide the student with an appropriate knowledge base and background for the practice of therapeutics (i.e., the clinical use of drugs to diagnose, prevent, treat, or cure disease) and to help the student assimilate the ever-changing body of knowledge about drugs. The objective of the graduate training programs (pre- and postdoctoral) was to provide individuals the opportunity to achieve a high degree of competence in pharmacological and physiological science, thus preparing them for teaching and research careers in this area of biomedical science. The objectives of the teaching program to students of the allied health professions and nursing students, as well as other undergraduate students, were to provide service courses for these groups and were similar to those for medical students. The service commitment to the university and community encompasses all parts of the above general mission.

The philosophy of the department was to create an atmosphere and environment conducive to the accomplishment of this mission. The department fostered research by recruiting and retaining outstanding faculty, postdoctoral fellows, and predoctoral trainees, providing space, resources, common equipment, core laboratories, and administrative support. All members of the faculty were expected to compete for and obtain sufficient extramural support to carry out independent research. The department also placed a high priority on quality teaching and realized that its educational mission set it apart from a research institute. Faculty were encouraged to improve their teaching skills through continuing education opportunities. In keeping with the university's Jesuit philosophy, service to others in the university and community continued to complement the department's research and teaching mission, and was strongly encouraged.

Organization and Governance of the Department

Aministrative Structure
The chair, Dr. Westfall, was the chief administrative officer of the department. Table 13 shows a flow chart of the administrative structure. The administrative offices were divided into two branches: fiscal services, which administered the financial operations of the department (grant management, salaries, orders, etc.), and operations, which took care of all other activities of the department, including manuscript and grant preparation, teaching activities, courses, seminars, etc. There was a senior administrator in charge of each branch: Joyce Jackson was fiscal services administrator and Mary Alice Kauling was operations administrator. The chair and these two senior administrators met on a weekly basis for the purpose of discussing all operations of the department office. Upon Ms. Jackson's retirement in 2004, Ms. Kauling took over both functions.

The department also had an executive committee consisting of five senior full professors. The purpose of the executive committee is described below.

Departmental Meetings
The department met on a monthly basis. At these meetings, departmental, school, and university activities were discussed and reports from various department- or university-wide committees were made. Several retreats of the department were also held for the purpose of discussing and evaluating teaching, research, and service functions. In addition, the department had a weekly journal club/colloquium series, a weekly seminar series, and a monthly evening colloquium (B&B). Faculty, postdocs, and predoctoral trainees usually attended these events. The teaching faculty participating in the various modules met regularly during the course to prepare and review exam questions and to discuss grades and any problems arising.

Standing Committees
There were several standing committees, including the Executive Committee, the Graduate Committee, and the Seminar Committee. In addition, there were section directors for the second-year graduate courses: Advanced and Selected Topics in Pharmacology and Physiology. These included sections on cell physiology,

TABLE 13 Organization and Administration, Department of Pharmacological and Physiological Science (as of 2003)

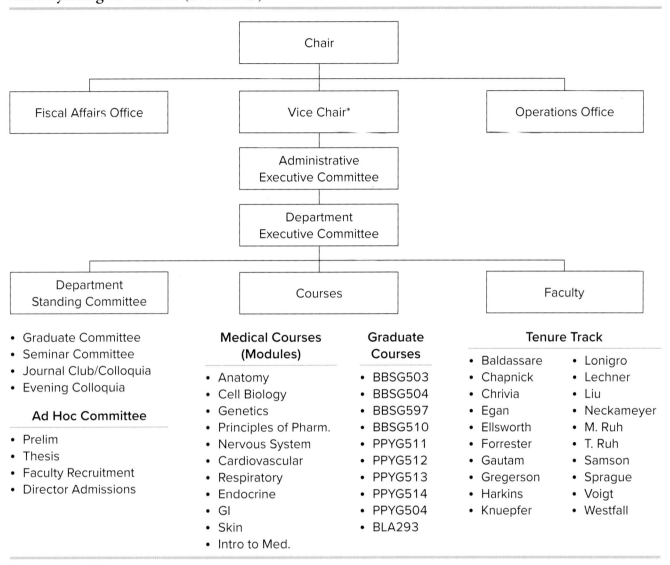

*Appointed in 2006

general pharmacological principles, the autonomic nervous system, the central nervous system, the cardiovascular/renal system, the endocrine system, and reproduction. There was also a course director for PPYG504 and 254, as well as a director of the journal club/colloquia and evening colloquia. In addition, there were numerous ad hoc committees, including the Faculty Recruitment and Search Committee and preliminary exam and thesis committees. The functions of some of these committees were as follows:

1. Vice Chair–A vice chair position was created in 2006. Major duties included advising the chair, representing him at important meetings, and making key decisions in his absence. Dr. Mark Voigt carried out this role from 2006 until January 2013, when he became interim chair for six months. He assumed the role of vice chair again from July 2013 until February 2018, when he once again became interim chair.
2. Executive Committee–The Executive Committee consisted of five senior full professors. Its purpose was to assist the chairman in making administrative decisions and other decisions affecting the department such as requests by faculty in other departments for secondary appointments in the department, recommendations to put younger faculty up for promotion, etc. This committee met approximately four to six times per year or when necessary. A member of the Executive Committee represented the chair at meetings occurring in his absence until the establishment of a vice chair in 2006.
3. Graduate Committee–This committee consisted of seven faculty members and one student member and was responsible for recruiting and selecting graduate students as well as supervising and monitoring graduate students' progress in the program. One member of the committee was director of graduate studies, and another was admissions director.

Ad Hoc Committees

4. From time to time, an ad hoc committee was formed. One of these was the Prelim Committee. When a graduate student was eligible to take his or her prelim exams, a prelim committee of five faculty was appointed to review the student's proposals and to conduct the oral exam in which the student defended his proposal. This committee was appointed by the Graduate Committee in consultation with the PhD advisor.
5. Dissertation Committee–Once a graduate student began his or her dissertation research, a dissertation committee was appointed. It consisted of the PhD advisor and at least three other faculty. This committee met at least twice a year to monitor the progress of the dissertation research and offer advice.
6. Faculty Search Committee–As departmental recruitments were instituted, a search committee was appointed for the purpose of evaluating potential new faculty.

Table 14 shows the names of course directors and chairs of the various committees.

Faculty Development Efforts

The following features were used to facilitate faculty development.

Setup Package to Facilitate Establishment of Research Laboratory

Each new faculty member was given a financial set-up package that was to be used to establish a functional and productive research laboratory. The funds provided could be used for equipment and supplies as well as for technical support. In addition, all core facilities and equipment were available to help in the development of an active and productive research program. Funds were provided for the faculty member's full salary until such time as he or she could also recover a proportion of salary from extramural support.

Mentoring by Senior Faculty and Chair

When a junior faculty member joined the department, two senior faculty members were assigned by the chair to act as mentors. The activities of these mentors, in addition to the chair, were to assist the new faculty member in identifying sources of grant support, help in preparing grant applications, assist the new faculty member in setting up his or her laboratory, give advice and help in teaching assignments, and evaluate teaching. The mentors met with the new faculty member periodically during the year to discuss his or her performance and success in the areas of research, teaching, and service. The mentors also gave advice as to how to respond to grant summary statements and reviews.

Meeting with the Chair
The chair met formally with each faculty member at least once a year, if not more often.

Discussions included status of research and scholarly productivity, teaching and service, and suggestions for continued development in these areas. For new faculty members, the meeting included a discussion of the requirements needed to obtain tenure and promotion.

Medical Student Evaluations
The Office of Curricular Affairs coordinated medical student evaluations of courses and faculty. These include written comments and scores on the teaching effectiveness of each faculty member participating in a respective course. When deemed appropriate, these were shared by the chair with individual faculty.

Peer Evaluation of Lecturers
Although the chair tried to attend as many lectures as he could in all the medical school modules and graduate courses, his time was limited. Therefore, the senior mentors described above were required to attend a sampling of lectures and provide appropriate feedback. For example, the organization and quality of PowerPoint presentations were critiqued and the syllabus was checked for clarity.

Midcycle Review
Assistant and associate professors underwent a midcycle review (generally carried out after three years for new assistant professors and three to four years for associate professors). These were conducted by the Internal Promotion and Tenure Committee. The findings and recommendations of the Internal Promotion and Tenure Committee were then transmitted to the chair. This same committee evaluated whether or not a faculty member should be proposed for promotion in rank and tenure (from assistant to associate professor) or promotion in rank (from associate to professor). The Internal Promotion and Tenure Committee consisted of four to five senior tenured professors appointed by the chair.

Weekly Seminar
Research seminars were scheduled on a weekly basis. Most speakers were from other institutions or departments. All faculty had the opportunity to suggest speakers who worked in their area of research. During the speakers' visits, faculty had the chance to discuss research and teaching strategies one on one.

Attendance at Scientific Meetings or Teaching Workshops
Some funds were available to supplement faculty travel to national and international scientific meetings. The chair encouraged faculty to attend teaching workshops at departmental expense.

Sabbatical Program
An active sabbatical program was in operation and all faculty were encouraged to take advantage of the opportunity to further develop and upgrade their skills. The chair informed the faculty member of the procedures to set up a sabbatical program.

Medical Education Activities

From the creation of the Department of Pharmacological and Physiological Science in 1990 until the curriculum reform of 1998, the department had the responsibility of teaching both the medical pharmacology course and the medical physiology course.

Medical Pharmacology Course
During this period, the medical pharmacology course remained essentially the same as that described previously for the Department of Pharmacology in Chapter Nine. The new courses formed as a result of the curriculum reform of 1998 will be discussed below.

Medical Physiology Course
With the creation of the Department of Pharmacological and Physiological Science in 1990, the department assumed responsibility for teaching the medical physiology course previously taught by

TABLE 14 Course Directors and Committee Chairs, Department of Pharmacological and Physiological Science (2001–02)

Principles of Pharmacology	Thomas C. Westfall
Cardiovascular Module	Mark Knuepfer
Respiratory Module	Andrew Lechner
General Physiology (PPYG504)	Thomas Ruh
Human Physiology (PPYG254)	Kevin Patton (later Joseph Baldassare and John Chrivia)
Advanced and Selected Topics in Physiology and Pharmacology (PPYG511, 512, 513, 514)	
• Director	Thomas C. Westfall
• Section on Cell Physiology	Terrance Egan
• Section on Pharmacological General Principles	Joseph Baldassare
• Section on Autonomic Nervous System	Thomas C. Westfall
• Section on Central Nervous System	Mark Voigt
• Section on Cardiovascular/Renal Respiration	Andrew Lonigro
• Section on Endocrine and Reproduction	Mary Ruh
Journal Club/Colloquia (PPYG691)	Barry Chapnick
Seminar Series (PPYG690)	Terrance Egan, John Chrivia, Medha Gautam
Evening Colloquia (Band B)	Al Stephenson
Graduate Committee:	
• Chair and Director	Barry Chapnick
• Admissions and Curriculum	Joseph Baldassare

the Department of Physiology. This was presented from 1990 to 1998, after which the new integrative curriculum replaced this course with various organ system modules. Faculty teaching this course came from the previous Departments of Physiology and Pharmacology.

Goals of the Physiology Course. The purpose of the course was to provide medical students with a solid introduction to the discipline of physiology. Physiology is the study of the normal function of tissues, organs, and systems of the body, and as such is universally recognized as one of the foundational subjects of medical education. It was the goal of the department to provide students with a series of lectures that emphasized the principal facts, discussed concepts of function and the underlying mechanisms that control the responses of tissues to stimulation, and provided up-to-date information not yet available in textbooks. Students were encouraged to supplement the lecture material by extensive readings from selected texts in order to have a comprehensive view of bodily functions. In addition, there were laboratory and problem-based exercises that reinforced the reading and lecture material and helped the integration of that information.

The course was scheduled from early November until early May in the first year's curriculum. It met four days a week, and four exams each represented 25 percent of the final grade. Table 15 shows a representative schedule and list of topics from the 1992–93 course. It is representative of all courses taught from 1990 until 1998.

Curricular Change and Reform

Based on the Liaison Committee on Medical Education (LCME) site visit and report of 1994, there was a major reorganization of the medical school curriculum, especially the first two years. The LCME report came down hard on the school for having too many lectures with emphasis on passive learning, rather than opportunities for the students to engage in independent study. This caused the administration and the Educational Policy Committee (consisting of representatives of all major courses) to form a committee to examine and evaluate the current curriculum. This was similar to what was going on nationally at many institutions and the advent of problem-based learning at many institutions. A year-long study took place during which each basic science course was examined in great detail. The outcome of these intensive examinations of the curriculum was a change from primarily department-based courses (e.g., anatomy, biochemistry, physiology, microbiology pharmacology, etc.) to an integrated curricular format. In the new format, no individual department was primarily in charge of or controlled a course; rather, these responsibilities transferred to a course director and the faculty identified to teach the module. The overall supervision of the module was the responsibility of the Office of Curricular Affairs. All courses became integrated, and an organ system/integrated curriculum model was adopted and began in 1998. During the first year of medical school, several new courses appeared, while in the second year a series of organ system modules were introduced. First year courses included Dissection (previously Gross Anatomy), Cell Biology, Metabolism, Genetics, Principles of Pharmacology, Introduction to Pathology, and Microbes and Host Responses. In the second year the organ system modules were grouped into neuroscience, respiratory, cardiovascular, renal/urinary, gastrointestinal, endocrine/reproduction, skin and connective tissue, and Introduction to Medicine.

The new curriculum had major implications for the Department of Pharmacological and Physiological Science, which had been solely responsible for the medical pharmacology and medical physiology course; both courses as such disappeared. All of the material that had been presented in the medical physiology course was incorporated into the new integrated course, including cell biology and the organ system modules studied by second year students.

For pharmacology, although the course as such disappeared, Dr. Westfall made a case that several components of the old pharmacology course did not fit well in the organ system model or approach.

These components were the general principles of drug action (absorption, distribution, biotransformation and excretion, or so-called pharmacokinetics) and the mechanism of drug action, or so-called pharmacodynamics, as well as drugs affecting the autonomic and somatic nervous systems, which involve all the organ systems; chemotherapy of parasitic, microbial, and neoplastic diseases; and toxicology, which likewise applies to all organ systems. These topics were incorporated into a new course entitled Principles of Pharmacology, which was placed in the first year.

Principles of Pharmacology Module

From its introduction in 1998 until the present time, this new course was primarily taught by faculty in the Department of Pharmacological and Physiological Science, although there were faculty from the Departments of Molecular Microbiology and Immunology, Internal Medicine, and Pediatrics who also participated. The course director was Dr. Westfall until his retirement in 2016, when Dr. Heather Macarthur was named course director. Dr. Joseph Baldassare was co-course director.

The Principles of pharmacology module consisted of lectures, case studies, review conferences, and practice exam questions, plus three innovative additions to the course: patient-oriented problem solving cases (POPS), computer exercises and tutorials identified on the web, and human simulation laboratories. Several POPS exercises were obtained from the pool of exercises originally developed by the Upjohn Pharmaceutical Company and taken over by the Association of Medical School Pharmacology Chairs (AMSPC). Human simulation labs were first introduced in the 2004 course. These were designed to illustrate the effect of autonomic drugs on cardiovascular parameters. The class was divided into twelve to fourteen groups and met at various times. (See pages 99–100 for an outline of these exercises). The Principles of Pharmacology module was held several times a week over a nine-week period from 1998 until 2013, for seven weeks from 2013 to 2015, and four to five weeks from 2015 to the present. The number of exams and quizzes changed through the years. From 1998 to 2011, there were two exams, each equaling 50 percent of the final grade. In the 2012–13 academic

TABLE 15 Lecture, Review, and Exam Schedule of Medical Physiology Course PY705 (1992–93)

Date	Day	Time	Topic	Lecturer
Section 1. Cellular and Neurophysiology				
Nov. 4	Wed	9:00-9:50	Greeting of Students and General Comments	Dr. Westfall
			Introduction to Physiology and Course Guidelines	Dr. Lechner
		10:00-10:50	Principles of Membrane Transport I	Dr. Ellsworth
Nov. 5	Thur	9:00-9:50	Principles of Membrane Transport II	Dr. Ellsworth
Nov. 9	Mon	9:00-9:50	Origin of the Resting Membrane Potential	Dr. Andrade
Nov. 10	Tue	9:00-9:50	Ionic Basis of the Action Potential I	Dr. Andrade
Nov. 11	Wed	9:00-9:50	Ionic Basis of the Action Potential II	Dr. Andrade
		10:00-10:50	Neuromuscular and Synaptic Transmission	Dr. Andrade
Nov. 12	Thur	9:00-9:50	Basic Concepts of Striated Muscle Activity	Dr. Ellsworth
Nov. 16	Mon	9:00-9:50	Basic Concepts of Smooth Muscle Activity	Dr. Ellsworth
Nov. 17	Tue	9:00-9:50	Properties of the Myocardium I	Dr. Forrester
Nov. 18	Wed	9:00-9:50	Properties of the Myocardium II	Dr. Forrester
Section 2. Autonomic Nervous System				
Nov. 18	Wed	10:00-10:50	Anatomical Considerations I	Dr. Westfall
Nov. 19	Thur	9:00-9:50	Anatomical Considerations II	Dr. Westfall
Nov. 23	Mon	9:00-9:50	Cholinergic Neurotransmission	Dr. Westfall
Nov. 24	Tue	9:00-9:50	Noradrenergic Neurotransmission	Dr. Westfall
Nov. 30	Mon	9:00-9:50	Physiological Considerations I	Dr. Westfall
Dec. 1	Tue	9:00-9:50	Physiological Considerations II	Dr. Westfall
Dec. 2	Wed	9:00-9:50	Physiological Considerations III	Dr. Westfall
		10:00-10:50	Physiological Considerations IV	Dr. Westfall
Section 3. Blood and Immunity				
Dec. 3	Thur	9:00-9:50	Erythrocytes and Erythropoiesis	Dr. Lechner
Dec. 7	Mon	9:00-9:50	Hemoglobin Metabolism; Erythrocyte Abnormalities	Dr. Lechner
Dec. 8	Tue	9:00-9:50	Hemoglobin Function: O_2 and CO_2 Transport	Dr. Lechner
Dec. 9	Wed	9:00-9:50	Granulocytes and Mononuclear Leukocytes	Dr. Lechner
		10:00-10:50	Blood Groups Antigens; Platelet Structure and Function	Dr. Lechner
Dec. 10	Thur	9:00-9:50	Secondary Hemostasis, Coagulation, Fibrinolysis	Dr. Lechner
Dec. 11	Fri	9:00-10:50	Review Conference for Sections 1-3	Sections Faculty
Dec. 16	Wed	9:00-12:00	PY705 Exam #1 (LRC Auditorium)	Sections Faculty
Section 4. Cardiovascular Physiology				
Jan. 4	Mon	9:00-9:50	Introduction to the Circulation	Dr. Forrester
		10:00-10:50	Overview of the Blood Vessels and Heart	Dr. Forrester
Jan. 5	Tue	9:00-9:50	Principles of Electrocardiography	Dr. Forrester
Jan. 6	Wed	9:00-9:50	The Cardiac Cycle I	Dr. Forrester
Jan. 7	Thur	9:00-9:50	The Cardiac Cycle II	Dr. Forrester
Jan. 11	Mon	9:00-9:50	Hemodynamics	Dr. Forrester
		10:00-10:50	Regulation of Cardiac Function	Dr. Forrester
Jan. 12	Tue	9:00-9:50	Control of Arterial Blood Pressure	Dr. Forrester
Jan. 13	Wed	9:00-9:50	Cardiovascular Reflexes	Dr. Forrester
Jan. 14	Thur	9:00-9:50	General Concepts of Peripheral Blood Flow	Dr. Ellsworth
Jan. 19	Tue	9:00-9:50	Regulation of Coronary Blood Flow	Dr. Forrester
Jan. 20	Wed	9:00-9:50	Special Circulations	Dr. Ellsworth

TABLE 15 Lecture, Review, and Exam Schedule of Medical Physiology Course PY705 (1992–93), *cont'd*

Date	Day	Time	Topic	Lecturer
Section 5. Respiratory Physiology				
Jan. 21	Thur	9:00-9:50	Functional Anatomy of Respiration	Dr. Lechner
Jan. 25	Mon	9:00-9:50	Mechanics of Ventilation; Principles of Spirometry	Dr. Lechner
		10:00-10:50	Lung Compliance; Obstructive and Restrictive Diseases	Dr. Lechner
Jan. 26	Tue	9:00-9:50	Regional Lung Ventilation; Pulmonary Circulation	Dr. Lechner
Jan. 27	Wed	9:00-9:50	Pulmonary Fluid Balance and Edema Formation	Dr. Lechner
Jan. 28	Thur	9:00-9:50	Gas Diffusion and Associated Physiological Equations	Dr. Lechner
Feb. 1	Mon	9:00-9:50	Gas Diffusion in Alveoli; Physiological Shunt	Dr. Lechner
		10:00-10:50	CO_2 Transport and Pulmonary Acid-Base Balance	Dr. Lechner
Feb. 2	Tue	9:00-9:50	Chemoreceptors and Control of Ventilation	Dr. Lechner
Feb. 3	Wed	9:00-9:50	Special Topics in Respiration	Dr. Lechner
Feb. 4	Thur	9:00-9:50	Review Conference for Sections 4-5	Sections Faculty
Feb. 11	Thur	1:00-4:00	PY705 Exam #2 (LRC Auditorium)	Sections Faculty
Section 6. Body Fluids, Renal Function, and Acid-Base Balance				
Feb. 22	Mon	9:00-9:50	Introduction–Body Fluid Compartments	Dr. Martin
Feb. 23	Tue	9:00-9:50	Structure-Function Relationship	Dr. Bastani
Feb. 24	Wed	9:00-9:50	Regulation of Glomerular Filtration Rate and Renal Blood Flow	Dr. Schmitz
Feb. 25	Thur	9:00-9:50	Transport Functions I–Clearance Concepts	Dr. Gellens
Feb. 26	Fri	9:00-9:50	Transport Functions II–Proximal Tubule	Dr. Gellens
Mar. 1	Mon	9:00-9:50	Transport Functions II–Distal Tubule	Dr. Schmitz
Mar. 2	Tue	9:00-9:50	Countercurrent Multiplication–Basis for Urinary Concentration and Dilution	Dr. Schmitz
Mar. 3	Wed	9:00-9:50	Renal Endocrinology	Dr. Martin
Mar. 4	Thur	9:00-9:50	Integration of Body Fluid Homeostasis	Dr. Martin
Mar. 5	Fri	9:00-9:50	Acid-Base Physiology	Dr. Moskowitz
Mar. 8	Mon	9:00-9:50	Renal Regulation of Acid-Base Homeostasis	Dr. Bastani
Mar. 9	Tue	9:00-9:50	Integration of Renal Physiology and Clinical Medicine	Dr. Moskowitz
Mar. 10	Wed	9:00-9:50	Review Conference for Section 6	Section Faculty
Section 7. Circulatory Shock				
Mar. 11	Thur	9:00-9:50	Pathophysiology of Shock I	Dr. Liu
Mar. 12	Fri	9:00-9:50	Pathophysiology of Shock II	Dr. Liu
Mar. 15	Mon	9:00-9:50	Compensatory Mechanisms and Management of Shock	Dr. Liu
Mar. 16	Tue	9:00-9:50	Open Date To be Announced	
Section 8. Thermoregulation and Exercise Physiology				
Mar. 17	Wed	9:00-9:50	Aerobic and Anaerobic Energy Production	Dr. Senay
Mar. 18	Thur	9:00-9:50	Energy Liberation during Rhythmic Exercise; Respiratory Exchange	Dr. Senay
Mar. 19	Fri	9:00-9:50	Circulation in Exercise; Isometric Exercise	Dr. Senay

TABLE 15 Lecture, Review, and Exam Schedule of Medical Physiology Course PY705 (1992–93), *cont'd*

Date	Day	Time	Topic	Lecturer
Section 8. Thermoregulation and Exercise Physiology, *cont'd*				
Mar. 22	Mon	9:00-9:50	Basic Thermoregulatory Values and the Heat Equation	Dr. Senay
Mar. 23	Tue	9:00-9:50	Heat Transfer Explained Using the Heat Equation	Dr. Senay
Mar. 24	Wed	9:00-9:50	Review Conference for Sections 7-8	Sections Faculty
Mar. 25	Thur	9:00-9:50	Open Date	To be Announced
Mar. 30	Tue	9:00-12:00	PY705 Exam #3 (LRC Auditorium)	Sections Faculty
Section 9. Endocrinology				
Apr. 5	Mon	9:00-9:50	General Concepts of Endocrinology	Dr. M. Ruh
Apr. 6	Tue	9:00-9:50	Hormone Binding Kinetics: Mechanisms of Steroid Hormone Action	Dr. T. Ruh
Apr. 7	Wed	9:00-9:50	Mechanisms of Peptide Hormone Action	Dr. Partridge
Apr. 8	Thur	9:00-9:50	The Hypothalamus	Dr. Partridge
Apr. 12	Mon	9:00-9:50	Anterior Pituitary	Dr. Beinfeld
Apr. 14	Tue	9:00-9:50	Posterior Pituitary	Dr. Partridge
Apr. 15	Wed	9:00-9:50	Adrenal Cortex	To be Announced
		10:00-10:50	The Thyroid	Dr. Partridge
Apr. 16	Thur	9:00-9:50	Calcium Homeostasis	Dr. Partridge
Apr. 19	Mon	9:00-9:50	Glucose Homeostasis	Dr. Partridge
Apr. 20	Tue	9:00-9:50	Male Reproduction	Dr. M. Ruh
Apr. 21	Wed	9:00-9:50	Female Reproduction I	Dr. M. Ruh
Apr. 22	Thur	9:00-9:50	Female Reproduction II	Dr. M. Ruh
Section 10. Gastrointestinal Physiology				
Apr. 23	Fri	9:00-9:50	Introduction and G.I. Hormones	Dr. M. Ruh
Apr. 26	Mon	9:00-9:50	Neuromuscular System	Dr. M. Ruh
Apr. 28	Wed	9:00-9:50	Vomiting, Gastric Emptying, and Secretion	Dr. M. Ruh
Apr. 29	Thur	9:00-9:50	Control of Gastric Acid Secretion	Dr. T. Ruh
Apr. 29		10:00-10:50	Ulcers, the Pancreas, the Liver, and the Gall Bladder	Dr. T. Ruh
Apr. 30	Fri	9:00-9:50	Small Intestinal Motility	Dr. M. Ruh
May 3	Mon	9:00-9:50	The Colon, Intestinal Secretions, Water and Electrolyte Absorption	Dr. M. Ruh
May 5	Wed	9:00-9:50	Absorption of Water and Electrolytes	Dr. M. Ruh
May 6	Thur	9:00-9:50	Digestion and Absorption of Carbohydrates and Proteins	Dr. M. Ruh
		10:00-10:50	Digestion and Absorption of Lipids	Dr. M. Ruh
May 7	Fri	9:00-9:50	Review Conference for Sections 9-10	Sections Faculty
May 13	Thur	1:00-5:00	PY705 Exam #4 (Final) (LRC Auditorium)	All PY705 Faculty

year, there were three exams, and in 2014, two quizzes and one cumulative final. Due to a reduction in hours assigned to the course, toxicology was rescheduled to a different course in 2015, and simulation labs were deleted in 2016 and 2017 but reintroduced in 2018. Table 16 depicts a representative schedule for the Principles of Pharmacology course taught in 2012. This is representative of courses taught from 1998 to 2016.

Course Objective of Principles of Pharmacology

Principles of Pharmacology course objectives were designed to lay the groundwork for the integrative study of the use of drugs for the diagnosis, prevention, treatment, and cure of disease. This information has a direct bearing on what is studied in the organ system modules during the following year when the effects of drugs in specific organ systems are studied. This information also provides the background necessary when students use drugs therapeutically in the clinic. The course consists of four sections:

- General Principles of Pharmacology
- Drugs Affecting Synaptic and Neuroeffector Junction–Autonomic and Somatic Nervous System
- Drugs Used in the Chemotherapy of Neoplastic, Microbial, Parasitic, and Viral Diseases
- General Principles of Toxicology

These sections were presented in Principles of Pharmacology because they do not lend themselves well to placement in specific organ system modules. Upon completion of this course, students should be able to describe and discuss the general principles that affect all therapeutic agents, that is, the absorption, distribution, biotransformation, excretion, mechanism of action, and toxic effects of drugs. In addition, the students should be able to describe and discuss the pharmacology of drugs affecting the autonomic and somatic nervous systems as well as chemotherapeutic agents. Regarding these agents, students should recognize the effects and actions of drugs on cells, tissues, organ systems, and the whole patient. In addition, students should be able to describe and discuss the mechanism of action of specific drugs at various levels of biological organization, the pharmacokinetics of these drugs, and their therapeutic uses and adverse effects.

Organization, Structure, and Examinations

Principles of Pharmacology consists of lectures, case studies in toxicology, assigned computer exercises and web tutorials, clinical correlations, and review conferences. There are generally three exams in the course. The first exam tests knowledge of material covered in Section 1 (General Principles of Pharmacology) and Section 2 (Drugs Affecting Synaptic and Neuroeffector Junction–Autonomic and Somatic Nervous System). This exam covers approximately 50 percent of the course. The second exam tests knowledge of the material covered in Section 3 (Chemotherapy of Neoplastic, Microbial, Parasitic, and Viral Disease). The third exam tests knowledge of the material covered in Section 4 (General Principles of Toxicology). Exams 2 and 3 each represent approximately 25 percent of the course. Practice questions are available weekly throughout the course. Both the practice questions and the three exams are of the best answer (National Board) format. Numerous practice exam questions are also included in the computer exercises. A feature of the course is human simulation exercises. These have been designed to illustrate the effect of autonomic drugs on cardiovascular parameters. Classes are typically divided into sixteen groups, with each group session lasting approximately one hour and fifteen minutes. Students are responsible for the material covered in lectures, tutorial cases, assigned computer exercises, and the human simulation exercises.

Grade Policy Guidelines

Students are graded on a three-tier system: Pass (P), Fail (F), Incomplete (I).

- Pass (P): Students who earn the grade of Pass (P) have successfully met the course requirements (greater than or equal to 65 percent).
- Fail (F): The grade of Fail (F) designates a student performance that does not meet minimum standards (less than 60 percent).
- Incomplete (I): The grade of Incomplete (I) designates a student performance of less than

64 percent and greater than 60 percent. This grade can be remediated by reexamination.

Physiology and Pharmacology of Drugs Acting on Specific Organ Systems

Because the course in medical physiology disappeared from the curriculum after the curriculum reform and reorganization in 1998, that material is now taught in a number of new courses and organ system modules as described in Table 18. Likewise, the actions of drugs affecting specific organ systems, and no longer taught in Principles of Pharmacology (Table 16), are taught in these organ system modules. Faculty from the Department of Pharmacological and Physiological Science are primarily responsible for teaching the physiology and pharmacology covered in these new courses. Table 18 shows the physiology and pharmacology topics in these various courses covered by faculty from the department. The entire course schedules are not included.

Graduate Training Activities

Graduate Program and Graduate Education

Upon the establishment of the Department of Pharmacological and Physiological Science in 1990, the graduate programs that existed in the Department of Pharmacology and the Department of Physiology were consolidated into the graduate program in Pharmacological and Physiological Science. The objective of this training program was to provide individuals with the opportunity to achieve a high degree of competence in the extensive area of pharmacological and physiological sciences, thus preparing them for research, academic, or alternate careers. The overall goals of the PhD program in the department were to instill enthusiasm for discovery and the scientific process, foster the development of critical thinking skills, develop laboratory research competence, develop oral and written communication skills, promote a commitment to scholarship. Based on the concept that pharmacological and physiological science is a broadly based subject, the PhD training program stressed these highly integrative and multidisciplinary aspects. For several years the requirements for the PhD in pharmacological and physiological science were similar to those for the PhD in pharmacology or physiology described previously. Basically, these were courses or competence in biochemistry (Biochemistry 501, 503), physiology (Physiology 506, 507), pharmacology (Pharmacology 501, 502), pharmacological and physiological journal club colloquia; pharmacological and physiological seminars; the three advanced courses in cardiovascular science, neuropharmacology, and endocrine/signal transduction; competency in statistics; ethical conduct of research; passing the qualifying/prelim exam; and successful completion of a dissertation research project.

Two major changes took place in 1997–98 that had a profound influence on the graduate training program in the Department of Pharmacological and Physiological Science. The first was the medical school curriculum evaluation and reform that took place that year, resulting in the adoption of the new integrated medical school curriculum described above. The second was the creation of the umbrella entry program called the Core Graduate Program in Biomedical Sciences.

Development of the Core Graduate Program in Biomedical Sciences (1998–Present)

Up until 1998, all the basic science departments at Saint Louis University had completely independent graduate programs. There were PhD programs in anatomy and neurobiology, biochemistry, microbiology, pharmacological and physiological science (two distinct programs prior to 1990), and pathology. In addition, there was one interdepartmental program in cell and molecular biology (CMB). Each of these programs did its own recruiting and had curriculum totally related to its own discipline. The CMB program had representatives from all the department-based programs and was administered by an interdisciplinary committee. During the middle and late 1990s, two faculty who were program directors of NIH training grants—Bill Longmore of the Department of Biochemistry and Molecular Biology, Program Director of a training grant in Cardiovascular and Pulmonary Science from NHLBI, and Tom Westfall, Department of Pharmacological and Physiological

TABLE 16 Principles of Pharmacology Course Schedule (2012)

Week 33

March 19, Monday			
10:00-11:50	Introduction to Course; Drug Development and History	Dr. Westfall	EU Aud
March 20, Tuesday			
1:00-3:50	Drug Absorption and Distribution	Dr. Baldassare	EU Aud
March 22, Thursday			
10:00-11:50	Biotransformation and Excretion	Dr. Westfall	EU Aud
March 23, Friday			
10:00-11:50	Factors Modifying Drug Action	Dr. Westfall	EU Aud
1:00-3:00	Introduction to Pharmacokinetic Principles	Dr. Baldassare	EU Aud
3:00-4:00	Pharmacokinetic Problems I	Dr. Baldassare	EU Aud

Week 34

March 26, Monday			
10:00-11:50	General Mechanisms of Drug Action	Dr. Baldassare	EU Aud
March 27, Tuesday			
1:00-3:00	Pharmacokinetic Problems II	Dr. Baldassare	EU Aud
March 28, Wednesday			
9:00-11:50	General Principles of Drugs Acting at Neuroeffector Junctions (ANS I)	Dr. Westfall	EU Aud
March 29, Thursday			
9:00-11:50	General Principles of Drugs Acting at Neuroeffector Junctions (ANS II)	Dr. Westfall	EU Aud
March 30, Friday			
9:00-11:50	Cholinergic Drugs I	Dr. Westfall	EU Aud
1:00-2:50	Geropharmacology	Dr. Flaherty	EU Aud

Week 35

April 2, Monday			
9:00-11:50	Cholinergic Drugs II / Adrenergic Drugs I	Dr. Westfall	EU Aud
April 5, Thursday			
9:00-11:50	Adrenergic Drugs II	Dr. Westfall	EU Aud
April 6, Friday	Good Friday	NO CLASS	

Week 36

April 9, Monday			
9:00-10:15	Human Sim. Labs (1-2)	Drs. Westfall/Macarthur	LOOI
10:30-11:45	Human Sim. Labs (3-4)	Drs. Westfall/Macarthur	LOOI
April 10, Tuesday			
9:00-10:15	Human Sim. Labs (5-6)	Drs. Westfall/Macarthur	LOOI
10:30-11:45	Human Sim. Labs (7-8)	Drs. Westfall/Macarthur	LOOI
April 11, Wednesday			
9:00-10:15	Human Sim. Labs (9-10)	Drs. Westfall/Macarthur	LOOI
10:30-11:45	Human Sim. Labs (11-12)	Drs. Westfall/Macarthur	LOOI
April 12, Thursday			
9:00-10:15	Human Sim. Labs (13-14)	Drs. Westfall/Macarthur	LOOI
10:30-11:45	Human Sim. Labs (15-16)	Drs. Westfall/Macarthur	LOOI

TABLE 16 Principles of Pharmacology Course Schedule (2012), *cont'd*

Week 37
April 16, Monday
   9:00-9:50   Adrenergic Drugs III   Dr. Westfall   EU Aud
   10:00-11:50   Review   Drs. Westfall/Baldassare   EU Aud
April 18, Wednesday   Study Day
April 19, Thursday
   9:00-12:00   Exam I   Pitlyk Auditoriums A, B, C

Week 38
April 23, Monday
   9:00-11:50   Cancer Chemotherapy   Dr. Chrivia   EU Aud
April 26, Thursday
   9:00-9:50   Introduction to Antimicrobial Chemotherapy   Dr. Westfall   EU Aud
   10:00-11:50   β-Lactam & Related Drugs   Dr. Enkvetchakul   EU Aud
April 27, Friday
   9:00-10:50   Aminoglycosides & Tetracyclines   Dr. Westfall   EU Aud
   11:00-11:50   Sulfonamides and Quinolones   Dr. Enkvetchakul   EU Aud
   1:00-2:50   Clinical Correlation Cancer   Dr. Pincus   EU Aud
   3:00-3:50   Protazoan   Dr. Chrivia   EU Aud

Week 39
April 30, Monday
   9:00-9:50   Miscellaneous Antibiotics   Dr. Chrivia   EU Aud
   10:00-11:50   Anti-Viral Drugs   Dr. Wold   EU Aud
May 2, Wednesday
   9:00-9:50   Anti-Helminthics   Dr. Chrivia   EU Aud
   10:00-10:50   Anti-Fungal Drugs   Dr. Enkvetchakul   EU Aud
   11:00-11:50   Antimycobacterial Agents   Dr. Chrivia   EU Aud
May 3, Thursday   Study Day
May 4, Friday
   9:00-10:50   Exam II   EU Aud

Week 40
May 7, Monday
   9:00-9:50   Toxicology I   Drs. Scalzo & Tominack   EU Aud
   10:00-10:50   Toxicology II   Drs. Scalzo & Tominack   EU Aud
   11:00-11:50   Toxicology III   Drs. Scalzo & Tominack   EU Aud
May 9, Wednesday
   9:00-10:00   Toxicology IV   Drs. Scalzo & Tominack   EU Aud
   10:00-10:50   Toxicology V   Drs. Scalzo & Tominack   EU Aud
   11:00-11:50   Toxicology VI   Drs. Scalzo & Tominack   EU Aud
   1:00-1:50   Toxicology VII   Drs. Scalzo & Tominack   EU Aud
May 10, Thursday
   10:00-10:50   Toxicology VIII   Drs. Scalzo & Tominack   LRC Aud A
   11:00-11:50   Toxicology IX   Drs. Scalzo & Tominack   LRC Aud A
May 11, Friday   Study Day

Week 41
May 15, Tuesday   Study Day
May 16, Wednesday
   9:00-10:50   Exam III   Pitlyk Auditorium A, B, C

Human Simulation Laboratory
Pharmacological Manipulations and Regulation of Arterial Blood Pressure
A CARDIOVASCULAR DEMONSTRATION UTILIZING THE HUMAN SIMULATOR

INTRODUCTION

As pointed out by Dr. Frank Vincenti in his virtual laboratory exercise, University of Washington, several generations of medical students learned practical aspects of autonomic and cardiovascular physiology and pharmacology from traditional exercises carried out on anesthetized dogs or cats. During these exercises, students would spend 3-4 hours injecting drugs, monitoring responses and discussing the various possible mechanisms of their observations. The vast majority of students thought these exercises were an extremely valuable learning experience.

For many reasons, we have modified these individual student exercises to the one being presented today which uses the human simulator rather than live animals. The session is a demonstration and discussion of various pharmacological manipulations of various cardiovascular parameters.

Experimental Manipulations

We will examine: 1) the effect of a variety of autonomic transmitters, hormones or drugs on the various cardiovascular parameters and 2) the effect of α-adrenoceptors or β-adrenoceptor antagonist.

Experimental Preparation and Parameters Measured

The Human Patient Simulator consists of a mannequin who has been anesthetized and a variety of parameters are being measured exactly as in the operating room. These include heart rate (beats/min), mean arterial blood pressure (mmHg), cardiac output (ℓ1/min) and

TABLE 17 Human Patient Simulator Worksheet

	Active Receptors	HR (bpm)	Art. Press (mm/Hg)	Mean Art. Press.	Card. Output (1/min)	Stroke Vol. *1000	Systemic Vascular Resistance
BASELINE		72-74	118/52	74	5.8-6.0	81	13
Norepinephrine 25 µg							
Epinephrine 25 µg							
Phenylephrine 150 µg							
Isoproterenol 10 µg							
Vasopressin 2 units							
Phentolamine 3 mg							
Phentolamine 3 mg + Norepinephrine 25 µg							
Phentolamine 3 mg + Phenylephrine 150 µg							
Phentolamine 3 mg + Vasopressin 2 units							
Propranolol 3 mg							
Propranolol 3 mg + Epinephrine 25 µg							
Propranolol 3 mg + Isoproterenol 10 µg							

Pharmacological Manipulation
Agonist Alone
After obtaining baseline data the following will be evaluated following IV bolus:
- norepinephrine (25 µg)
- epinephrine (25 µg)
- phenylephrine (150 µg)
- isoproterenol (10 µg)
- vasopressin (2 units)

Agonist after phentolamine
- phentolamine (3 mg)
- phentolamine (3 mg) + norepinephrine (25 µg)
- phentolamine (3mg) + phenylephrine (150 µg)
- phentolamine (3 mg) + vasopressin (2 Units)

Agonist after propranolol
- propranolol (3 mg)
- propranolol (3 mg) + norepinephrine (25 µg)
- propranolol (3 mg) + isoproterenol (10 µg)

Program Director of a T32 in Pharmacological Sciences from NIGMS—were both serving on NIH training grant review committees. They began talking about a major national trend: the appearance of umbrella graduate programs. These programs took one of two forms. One was an umbrella program for recruiting and for the initial phase of graduate training, followed by trainees earning the actual PhD in a department-based program. The second was an umbrella program for the entire graduate training and an undifferentiated PhD in biomedical science. The big advantage of both types of programs appeared to be the recruitment of better PhD candidates and training in topics applicable to all PhD trainees. Dr. Longmore and Dr. Westfall thought that an umbrella program for recruitment and for the first year of the graduate program deserved consideration. Numerous meetings were held with the various PhD program directors and chairs of the departments. After much discussion, it was agreed that such an umbrella program be adopted, and the Graduate Program in Biomedical Sciences, or the "core program," as it became known, was christened in 1998. Up until this time, each of the department-based programs received a number of funded slots from the school of medicine to support their graduate programs (the total number of slots was thirty-seven). It was agreed by all chairs that these individual allotted slots would be given up to support the core program. It was also agreed together with the Dean of the School of Medicine that a permanent full-time director be recruited to oversee the core program, with the advice of an Executive Committee consisting of departmental chairs or program directors of the individual department-based programs.

The Graduate Program in Biomedical Sciences was designed to provide students with a strong foundation in all aspects of basic biomedical science and freedom to explore diverse research opportunities during the first year of graduate training. The first-year curriculum combined lectures, small group discussion sessions, and journal clubs to develop self-confidence and familiarity with a breadth of biomedical science and technology that spanned the disciplines of biochemical, cell, molecular, developmental, genetic, neurobiological, and pharmacological science. At the end of the integrated first-year program, students would select a mentor and enter into one of the existing graduate programs of their choice. The students would ultimately earn their PhD in one of these programs: biochemistry and molecular biology, cell and molecular biology, molecular microbiology and immunology, pharmacological and physiological science, or pathology. Because of the reorganization of the Department of Anatomy and Neurobiology into the Center for Anatomical Education, the PhD

TABLE 18 Lecture Topics: Medical School Modules

Cell Biology Module
- Functional Organization of the Body — Dr. Sprague
- Signal Transduction I, II, III — Dr. Baldassare
- Transcription I and II — Dr. M. Ruh
- Membrane Transport and Homeostasis — Dr. Ellsworth
- Autonomic Nervous System I, II, III — Dr. Westfall
- Resting Membrane Potential — Dr. Egan
- Action Potential — Dr. Egan
- Synaptic Transmission — Dr. Egan
- Muscle Physiology I and II — Dr. Ellsworth
- How to Build a Blood Vessel and How Does It Work I and II — Dr. Sprague

Nervous System Module
- Synaptic Transmission Symposium — Drs. Ariel, Zahm, and Westfall
- Physiology of Reflexes and Proprioception — Dr. Forrester
- Physiology of Basal Ganglia and Cerebellum — Dr. Westfall
- Anti-Parkinson Drugs — Dr. Westfall
- Physiology of Somatosensory Receptors — Dr. Forrester
- Physiology of Proprioception — Dr. Knuepfer
- Physiology of Pain — Dr. Knuepfer
- Opioid Analgesics — Dr. Voigt
- Physiology of the Vestibular System — Dr. Ariel
- Physiology of the Auditory System — Dr. Forrester
- Visual System/Anatomy of the Eye and Retina — Dr. Ariel
- Ocular Motor System — Dr. Ariel
- Visual System/Physiology of the Eye — Dr. Ariel
- Visual System/Physiology of the Retina Targets and Cortical Processing — Dr. Ariel
- Cerebral Cortex — Dr. Ariel
- Physiology of Cerebral Blood Flow — Dr. Forrester
- Olfactory and Limbic System — Dr. Zahm
- Anatomy of the Hypothalamus — Dr. Samson
- Physiology of the Hypothalamus — Dr. Samson
- Taste and Autonomics — Dr. Samson
- Anti-Epileptic Drugs — Dr. Willmore
- Anti-Psychotic Drugs — Dr. Voigt
- Anti-Depressant Drugs — Dr. Voigt
- Anti-Anxiety Drugs — Dr. Voigt
- Sedative-Hypnotic Drugs — Dr. Westfall

Cardiovascular Module
- Introduction — Dr. Knuepfer
- Overview of Circulation — Dr. Forrester
- Autonomic Review — Dr. Westfall
- Blood Vessel Dynamics — Dr. Ellsworth
- Vasoactive Mediators I — Dr. Westfall
- Vasoactive Mediators II — Dr. Macarthur
- Blood Pressure Regulation — Dr. Knuepfer
- Physiology of the Heart — Dr. Herrmann
- Pathophysiology of the Heart — Dr. Herrmann
- Congestive Heart Failure — Dr. Herrmann
- Coronary Artery Pathophysiology — Dr. Herrmann
- Cardiac Ion Channels — Dr. Egan
- Antiarrhythmic Drugs — Dr. Knuepfer
- Pig Lab I, II, III — Drs. Knuepfer, Westfall
- Pig Lab Review — Dr. Knuepfer
- Antianginal Drugs — Dr. Westfall
- Antihyperlipidemics — Dr. Stephenson

Renal Module
- Anti-hypertensive Drugs I, II — Dr. Westfall

Respiratory Module
- Introduction — Dr. Lechner
- Red Cell Physiology — Dr. Lechner
- Spirometry and Mechanics of Normal Lung Ventilation — Dr. Lechner
- Pulmonary Defense Mechanisms Upper Airways — Dr. Sprague
- Integrated Lung Compliance and the Work of Breathing — Dr. Lechner
- Pulmonary Circulation, Capillary Filtration, Edema Formation — Dr. Lechner
- Ventilation/Perfusion Ratios — Dr. Lechner
- Calculating Critical Physiological Indicas of Lung Function — Dr. Lechner
- Alveolar O_2 Exchange & Physiological Shunt — Dr. Lechner
- Alveolar O_2 Exchange: Pulmonary Acid Base Balance — Dr. Lechner
- Case: Assessment of the Hypoxemic Patient — Dr. Sprague
- Case: Shortness of Breath — Dr. Sprague
- Control of Vascular Resistance — Dr. Sprague
- Pathophysiology of Vascular Resistance in the Lung — Dr. Sprague
- Pig Lab I, II, III — Drs. Lechner & Sprague

TABLE 18 Lecture Topics: Medical School Modules, *cont'd*

Skin, Connective Tissue, and Musculoskeletal Module		• Glucose Homeostasis	Dr. Samson
		• Pharmacology of Diabetes	Dr. Voigt
• Pharmacology of Anti Rheumatic Drugs	Dr. Stephenson	• Parathyroid and Calcium Homeostasis I, II	Dr. M. Ruh
• Pharmacology/Skin Therapy	Dr. Stephenson	• Adrenal Hormones I and II	Dr. Samson
		• Male Reproduction I and II	Dr. M. Ruh
Endocrine/Reproductive Module		• Female Reproduction I and II	Dr. M. Ruh
• Physiology/Pharmacology of Hypothalamus and Anterior/ Posterior Pituitary I, II	Dr. Samson		

program in anatomy was excluded. The core curriculum established for the core program was developed and taught by representatives of all the graduate programs.

In addition to the core program, PhD candidates with an advanced degree (e.g., a master of science) or advanced standing in a PhD program at another institution could be directly admitted into one of the department-based programs. MD/PhD students who completed their first or second year of medical school could also directly enter the departmental program of their choice.

Since 1998, most PhD students have entered graduate school via the core program. In 1999, Dr. Willis "Rick" Samson was hired as the permanent director of the core program. Dr. Samson reports directly to the Dean of the School of Medicine and is advised by an executive committee. During the first year of the core program's existence (1998–99), the program was directed by Dr. Nicola Partridge and Dr. Peggy Weideman. Programs currently participating in the core program are the PhD program in biochemistry and molecular biology, molecular microbiology and immunology, pathology, and physiology/pharmacology. The former PhD program in cell and molecular biology was disbanded in the early 2000s for lack of interest and redundancy.

Current Components of the Graduate Program

After successful completion of the one-year Graduate Program in Biomedical Sciences or direct admission (because of advanced standing), students enter the graduate program in pharmacological and physiological science. MD/PhD students who select pharmacological and physiological science for their graduate studies also enter the program (generally after their second year in medical school). After entry, each student selects a dissertation research advisor. The twenty-four credit hours of basic biomedical science courses, advanced standing (generally with an MD), or two years of medical school will have replaced most of the first-year course requirements formerly required by the pharmacological and physiological science program. Students entering from the core program or those with advanced standing will take an additional twelve credit hours in Advanced and Selected Topics in Pharmacology and Physiology (PPY 511, 512, 513, and 514, described on the following page and in Table 21), while MD/PhD students take selected components. In addition, all students take the journal club in Pharmacological and Physiological Science (PPY691) and seminars in Pharmacological and Physiological Science (PPY680). Students will also begin research on their dissertation project under the direction of their faculty advisor, and select a dissertation committee, and identify a proposition defense committee. See details and timeline below. The third year of graduate study and beyond is devoted almost exclusively to research related to the student's dissertation problem. It is required that the student meet with his or her dissertation committee at least twice a year. Throughout their program, students are required to participate in the departmental journal club and are encouraged to participate in specialized journal clubs, the weekly departmental seminar series, and the monthly evening colloquium as effective means of keeping up with new developments in science. Students may also elect to take advanced courses of their choosing from among the numerous offerings at Saint Louis University, Washington University, or the University of Missouri, St. Louis. The department requires PhD students to participate in teaching activities as part of their educational experience. These include Drugs We Use and Abuse (BLA245) and Human Physiology (PPY254)—both taught to undergraduate students at Saint Louis University. Students may also participate as tutors to medical students or students in the allied health program. Upon completion of the research work, students defend their work before their dissertation committee and present a public seminar on their work. Completion of the program, from entry into the core program to award of a PhD, usually requires five years of continuous study. A representative advertisement of the Graduate Program and training grant in Pharmacological and Physiological Science appears in Figure 27.

1. **Advanced Topics in Pharmacological and Physiological Sciences, PPY511 and PPY513.** These are four-credit courses that meet Monday, Tuesday, Thursday, and Friday from 9:00 to 11:00 a.m. The objectives are to expand the breadth of knowledge of mechanisms of drug action and functional and integrative biology. These courses will build upon and amplify the knowledge base presented in the first year basic biomedical science program. They integrate both pharmacological and physiological information into unified courses specifically designed for graduate students and replace the traditional pharmacology and physiology course that students formerly took with medical students. The courses are taught in modular form to facilitate the participation of students in programs other than the traditional pharmacological and physiological training program.

    PPY511: Advanced Topics in Pharmacological and Physiological Science I (4). This is an intense course integrating aspects of both physiology and pharmacology. Topics covered include general principles of cell physiology; general principles of pharmacology; pharmacology of the peripheral (autonomic, somatic) nervous system; and physiology and pharmacology of cardiovascular, renal, respiratory, and gastrointestinal systems. Meets in fall semester.

    PPY513: Advanced Topics in Pharmacological and Physiological Science II (4). An intense course integrating additional topics in physiology and pharmacology. Topics covered include physiology and pharmacology of the central nervous system, chemotherapy, endocrine systems, and male and female reproduction. Meets in spring semester.

2. **Selected Topics in Pharmacological and Physiological Sciences, PPY512 and PPY514.** These are two-semester, two-credit courses. The course grades are composed of a comprehensive final at the end of each semester. The comprehensive final is developed by the faculty that has taught during the relevant semester. A passing grade in both courses is required for eligibility to take the departmental preliminary examination for doctoral candidacy.

3. **Pharmacological and Physiological Science Seminar, PPY690 (0–1).** This course is scheduled weekly during the fall and spring semesters. Research seminars are presented by faculty and investigators from other departments of the university or guest speakers from other institutions. Approximately twenty to thirty outside speakers visit the department each year. Roundtable luncheon discussions with students and the speaker are regularly scheduled.

4. **Pharmacological and Physiological Science Journal Club, PPY691 (0–1).** A continuous weekly journal club/colloquium in which students and postdoctoral fellows discuss recent research findings and papers from the literature. Each student is required to present one journal club per year. The objectives, format, and evaluation procedure are similar to those applied for BBS597 and 598. The results of the faculty evaluation forms are discussed with each student individually by the journal club course director.

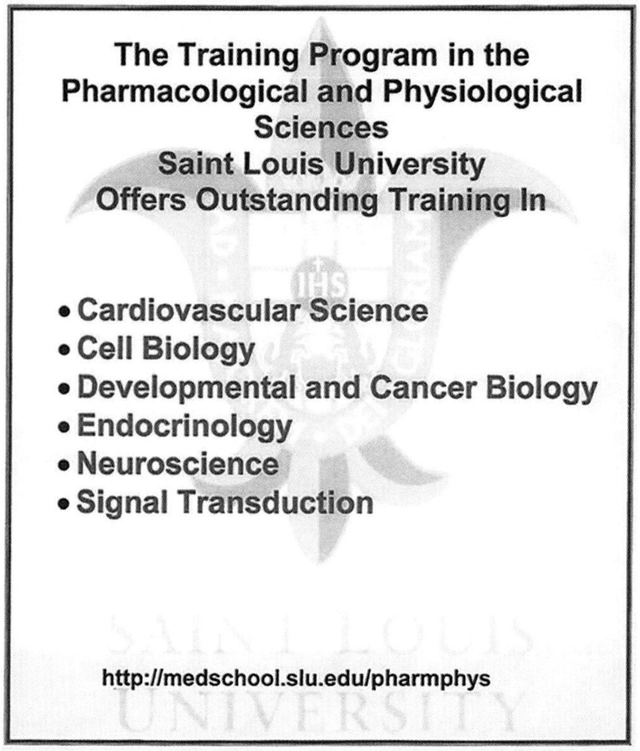

5. **Competency in Statistics.** This may be satisfied by a statistics course taken prior to entry in the program or by a formal course taken at Saint Louis University or elsewhere. The recommended course is Statistics for Health Sciences (LMH510-02), which has been specially designed for the needs of students in the basic medical sciences.

6. **Responsible Conduct of Research.** Training in the responsible conduct of research is required of all PhD students in the pharmacological and physiological science graduate program. This takes place in three phases:

    Phase 1. All students take and complete the web-based course CITI Biomedical Sciences Responsible Conduct of Research for the Unaffiliated Learner. This course consists of fourteen modules that cover seven content areas (research misconduct, data and management, responsible authorship, peer review, mentoring, conflicts of interest, and collaborations). After an introduction to each module, the student reviews additional material including case studies and completes a quiz. When all fourteen modules are successfully completed, the web-based program generates a certificate of completion. The director of the core graduate program (Dr. Willis Samson) collects copies of the certificates to ensure student compliance. For all students in the pharmacological training program, we require that the course be completed by the end of their first year in the core program and prior to joining the pharmacological science training program.

    Phase 2. Phase 2 is completed in the first year that students matriculate in the pharmacological science training program. All students are required to participate in a series of small group discussions covering several topics in the responsible conduct of research as part of our ten-month course PPYG 511/PPYG 513, Advanced Topics in Pharmacological and Physiological Sciences. The intent of these tutorials is to build on the principles learned in the core course by including mentor-led group discussions on a range of relevant topics. Each topic consists of an introductory one-hour didactic lecture, two to three case studies, and a small group discussion. Table 19 shows the topics to be covered

FIG. 27 Advertisement of training program in the Department of Pharmacological and Physiological Science

and the instructors who give the introductory lectures.

Phase 3. Phase 3 consists of discussion of various topics pertinent to the responsible conduct of research by individual training faculty in their laboratory meetings. Documentation that appropriate topics have been covered in these lab meetings will be provided by students during their biannual meeting with the graduate committee.

7. **Evening Colloquia.** Students also attend a monthly informal evening colloquium called B&B (beer and bull), in which one faculty member talks and another faculty member hosts at his or her home and provides refreshments. The purpose is to present ongoing and projected research. PowerPoint slides are not allowed, although the use of a chalkboard/easel is permitted. The format is evocative and informal. Any type of question is permitted, especially those that question the approach and strategy that underlies the research plan. The experience provides students with

insight into developing research strategies. These sessions and the faculty talks also help students become familiar with the types of research that are ongoing in various faculty laboratories.

Preliminary Examination for Advancement to Doctoral Candidacy

Overview of Preliminary Examination. Each student in the Department of Pharmacological and Physiological Science must successfully complete a written and oral examination in order to continue in the PhD program and advance to doctoral candidacy. The preliminary examination has three components. The first is a short written proposal that describes the rationale of the project, including the hypothesis and specific aims. The second component is a longer written proposal that follows the guidelines for the research proposal portion of an NIH R01 grant application. This includes background, any preliminary data (including that found in the literature), and experimental design of proposed experiments. The third component is an oral examination/defense of the written proposal. The proposal may be based on either the expected dissertation research project or a topic of research that is totally unrelated to that of the dissertation.

All MD/PhD students or students entering the program with advanced standing must also successfully complete the preliminary examination within the time frame described for traditional students. This requirement must be completed during the year of entrance to the program or the following year, depending on completion of coursework requirements.

Dissertation Committee and Defense. The purpose of the dissertation committee is to approve, advise, and evaluate research progress and make recommendations on the suitability for submission of the dissertation for defense. The dissertation committee should meet with the student at least twice a year. The PhD advisor will chair the dissertation committee, and it should be composed of at least two additional department members, depending on their expertise. In addition, a member of the committee may be appointed from outside the department. Upon approval by the dissertation committee, the student may write his or her dissertation. The public oral defense of the dissertation will be conducted as a formal seminar. The presentation will last approximately forty-five minutes and will be followed by a discussion/examination period, at which time all members of the audience, including the dissertation committee, may examine the PhD candidate. Spontaneous questions that arise during the presentation will not be discouraged. Immediately following the public examination, the members of the dissertation committee will meet privately with the candidate before the ballots are cast. The candidate will then be excused from the meeting in order for the committee to discuss its evaluations of the candidate and for committee members to cast their ballots.

TABLE 19 Responsible Conduct of Research, Phase 2

Topic	Title	Instructor
1	Ethical Use of Animals	Dr. John Long, DVM, Chair, Dept. of Comparative Medicine
2	Research Involving Human Subjects	James Willmore, MD, Professor of Neurology, Psychiatry, and Pharmacological and Physiological Science
3	Authorship: Guidelines and Ethical Issues	Willis Samson, PhD, Professor, Dept. of Pharmacological and Physiological Science, Director, Core Graduate Program in Biomedical Sciences
4	Mentorship and Traineeship	Thomas C. Westfall, PhD, Professor and Chair, Dept. of Pharmacological and Physiological Science, Program Director, Training Grant in Pharmacological Sciences
5	Conflict of Interest	Joseph Baldassare, PhD, Professor, Dept. of Pharmacological and Physiological Science, Director of Graduate Studies
6	Dealing with Scientific Misconduct	Terrance Egan, PhD, Professor, Dept. of Pharmacological and Physiological Science, Co-Director, Training Grant in Pharmacological Sciences

The ballots will be presented to the advisor with the final copies of the dissertation. If the committee requires minor revisions of the dissertation following the defense, all ballots will be withheld until every committee member is ready to cast their ballot. A unanimous positive evaluation of the dissertation committee, that is, all members whose signatures appear on the candidate's approved prospectus, is necessary for final approval of the dissertation. Should the candidate not be approved for graduation because of one negative vote from a dissertation committee member, the candidate may appeal. The appeal process is described in the catalog of the graduate school.

Teaching Opportunities

Graduate students in the program in pharmacological and physiological science or the training program in pharmacological sciences are required to obtain formal teaching experience by participating as lecturers in BLA245: Drugs We Use and Abuse. This is an undergraduate course for non-science majors presented in the fall semester. It carries three credits and meets Monday, Wednesday, and Friday from 10:00 to 11:00 a.m. Typical enrollment is seventy-five students. A state-of-the-art syllabus is prepared, with each chapter written by the lecturer. There are thirty-five lectures on different drugs. Last semester, thirteen advanced students participated and gave these lectures.

Students may also participate in Human Physiology (PPY254), which is an undergraduate course carrying four hours of credit. This is a fall semester course taken by two- to three hundred undergraduate health science students majoring in allied health sciences, nursing, psychology, or biology.

In addition to these two formal courses, graduate students have participated as tutors for medical students, students in the allied health professions program, or in the supplemental instruction program sponsored by the Office of Multicultural Affairs.

Monitoring of Student Progress and Academic Standards

During the year that students are in the Graduate Program in Biomedical Sciences, student progress is monitored by the director of the core program in consultation with the Curriculum Committee and the teaching faculty. During the second year (first year in the pharmacological and physiological training program), student progress is monitored by the Graduate Committee. The program director and director of graduate studies both maintain records of student grades, journal club presentations, progress on proposals, and selection of a thesis advisor. These are reviewed by the full committee semiannually, or more frequently if problems arise. The Graduate Committee meets with all students individually twice per year in the absence of the mentor to discuss students' progress and whether any problems exist. These meetings have been very successful in preventing or eliminating problems and keeping all students on track. During these meetings, students have the opportunity to provide information on their impressions of strengths and weaknesses of the training program and any recommendations for changes or improvement.

The required graduate school academic standard for all students receiving fellowships is the maintenance of a B average (3.0 grade point average) in each academic year. A student with a grade point average below 3.0 is normally put on academic probation and may lose the fellowship if this average is not restored to the 3.0 level in the next semester. At the time of the prelim defense, the thesis advisor reviews the progress with the rest of the student's proposition defense committee. From this point on, the primary monitoring of the students is done by the advisor and thesis committee. Table 20 depicts a representative curriculum for the graduate program.

Table 21 shows a representative schedule and list of topics for the PPY511–514 course taken from the Fall 2012 and Spring 2007 semesters.

Service Course Activities

The Department of Pharmacological and Physiological Science had responsibility for teaching several courses to nursing, allied health, and undergraduate students, as well as one group of post-baccalaureate students. These courses included Human Physiology (PPY254), General Physiology (PPYC504), Advanced Pharmacology in Primary Health Care Nursing (NAN508), Clinical Pharmacology–Physician Assistants (BLA245), Drugs We Use and Abuse;

and Human Systems Physiology in the Medical Anatomy and Physiology Program (MAPP). Some of these courses were taught only for a limited period of time, while others were taught during the entire period from 1990 to 2016. The origin of some of these courses is unclear and predated the Westfall era (e.g., PPY254, PPYG504), while others were established during the Westfall era (NRN 508, Clinical Pharmacology for Physician Assistants, the MAPP physiology course, and BLA245). Representative schedules for these various courses are depicted in Tables 22–26.

Drugs We Use and Abuse: BLA245

This course was established by Dr. Westfall to fulfill two major goals. First, it was thought that all liberal arts students should be exposed to some knowledge about drugs since they are such an important aspect of society. Secondly, such a course would provide the opportunity for graduate students in the pharmacological and physiological science program to obtain training and experience in formal teaching. BLA245 was established with the cooperation of the Department of Biology and carries a biology designation. It is an introductory, one-semester course for non-science majors that focuses on how drugs (legal and illegal) work and how they affect the function of the brain and body. The course is entirely taught by advanced graduate students who are working on PhDs in pharmacology and physiology. It consists of lectures, student presentations, review sessions, and examinations. It is taught from 10:00–10:50 a.m., Monday, Wednesday, and Friday for one semester and carries four credit hours. The textbook for the course is *Drugs, Society, and Human Behavior* by Ksir, Ray, and Hart.

Faculty Research Interests

Modern research in pharmacological and physiological science applies the experimental methods derived from clinical medicine, to biophysics, biochemistry, and molecular biology to provide a detailed understanding of drug action and physiological function at many different levels of inquiry. The Department of Pharmacological and Physiological Science is a major participant in the current excitement involving research in modern biomedical science. The faculty in the department have diverse backgrounds in the fields of biochemistry, medicine, molecular biology, neuroscience, pharmacology, and physiology. A common theme involves the physiology, pathophysiology, and pharmacology of intra- and intercellular communication. Major areas of specialization include neurotransmitter biochemistry, physiology, and pharmacology; the molecular biology, biochemistry, and pharmacology of neurotransmitter, autacoid, neurohumoral, and hormone receptors; intracellular signaling and transduction mechanisms; electrophysiology and ion channels; neurochemistry; cardiovascular and circulatory control mechanisms; the regulation and function of the autonomic, somatic, and central nervous systems; molecular, cellular, and endocrine control mechanisms; respiratory physiology; neuropharmacology; and drugs of abuse.

A summary of the research areas of interest for faculty who were in the department for some period between 1990 and 2013 (e.g., during the period that Dr. Westfall was chair) follows. A further synopsis of each faculty member's training and accomplishments appears later.

Full-Time Tenure-Track Faculty

- **Rodrigo Andrade, PhD, Yale University**
  Cellular and molecular neurobiology, electrophysiology, and serotonin neurotransmission in hippocampus and cortex.

- **Michael Ariel, PhD, Washington University**
  Neuroanatomy and neurophysiology of the ocular system.

- **Joseph Baldassare, PhD, University of Pittsburgh**
  Biochemical and molecular regulatory processes, intracellular transduction mechanism, cell cycle.

- **Margery Beinfeld, PhD, Washington University**
  Biochemical and molecular neuropharmacology, regulation of neuropeptide expression, processing and release, cholecystokinin.

Text continues on p. 117

TABLE 20 Representative Curriculum for the Graduate Program

Year 1: Core Basic Biomedical Science Curriculum

1st Semester	Credit Hours	Course Title
BBS501	5 hrs	• Basic Biomedical Sciences I
BBS502	4 hrs	• Special Topics in Basic Biomedical Sciences I
BBS597	2 hrs	• Introduction to Basic Biomedical Research
BBS592	1 hr	• Basic Biomedical Science Colloquium

2nd Semester	Credit Hours	Course Title
BBS503	5 hrs	• Basic Biomedical Sciences II
BBS504	4 hrs	• Special Topics in Basic Biomedical Sciences II
BBS597	2 hrs	• Introduction to Basic Biomedical Research
BBS592	1 hr	• Basic Biomedical Science Colloquium
BBS510	0 hr	• Ethics for Research Scientists
		• Informatics for Research Scientists

Summer Semester	Selection of a mentor and research project. Entry into one of the six individual graduate biomedical science programs.

Year 2: First Year of the Pharmacological and Physiological Sciences Program

1st Semester	Credit Hours	Course Title
PPY511	4 hrs	• Advanced Topics in Pharmacological and Physiological Sciences I
PPY512	2 hrs	• Selected Topics in Pharmacological and Physiological Science I
PPY691	1 hr	• Pharmacological and Physiological Science Journal Club
PPY680	1 hr	• Pharmacological and Physiological Science Seminar

2nd Semester	Credit Hours	Course Title
PPY513	4 hrs	• Advanced Topics in Pharmacological and Physiological Sciences II
PPY514	2 hrs	• Selected Topics in Pharmacological and Physiological Sciences II
PPY691	1 hr	• Pharmacological and Physiological Science Journal Club/Colloquia
PPY680	1 hr	• Pharmacological and Pharmacological Science Seminar

April-Summer	Prepare and Defend Preliminary Qualifying Examination for Advancement of Doctoral Candidates

Years 3; 4; 5: Pharmacological and Physiological Sciences Program

Title	Credit Hours	Title
• Directed Reading	0–3 hrs	• Competency in Statistics
• Research Topics	0–3 hrs	• Journal Club/Colloquia
• Electives	variable	• Seminar
• Dissertation Research	0–12 hrs	• Defense of Thesis

TABLE 21 Advanced and Selected Topics in Pharmacological and Physiological Science: PPY511–514

PPY511/512 taken from Fall 2012

9:00-11:00 a.m. Room M360

Date	Lecture Title	Faculty
Section #1: Fundamentals of Drug Action		
August 20	Pharmacokinetics I	Dr. Baldassare
21	Pharmacokinetics II	Dr. Baldassare
23	Drug Biotransformation	Dr. Westfall
24	Binding Theory I	Drs. Baldassare/Egan
27	Binding Theory II	Drs. Baldassare/Egan
28	Binding Theory PBL I	Drs. Baldassare/Egan
30	Efficacy and Potency	Drs. Baldassare/Egan
31	Efficacy and Potency PBL	Drs. Baldassare/Egan
September 3	Labor Day Holiday	
4	Partial Agonist and Antagonists	Drs. Baldassare/Egan
6	Partial Agonists PBL	Drs. Baldassare/Egan
7	Factors Modifying Drug Action	Dr. Westfall
10	Practical Measures of Drug Effects	Drs. Baldassare/Egan
11	Drug Design and Development	Dr. Westfall
13	Study Day	
14	Exam 1	
Section #2: The Physiology and Pharmacology of the Autonomic and Somatic Nervous System		
September 17	Synaptic and Neuroeffector Junctions I	Dr. Westfall
18	Synaptic and Neuroeffector Junctions II	Dr. Westfall
20	Cholinergic Agonists and Antagonists I	Dr. Westfall
21	Cholinergic Agonists and Antagonists II	Dr. Westfall
24	Adrenergic Agonists and Antagonists I	Dr. Westfall
25	Adrenergic Agonists and Antagonists II	Dr. Westfall
27	Wrap Up	Dr. Westfall
28	Exam 2	
Section #3: The Physiology and Pharmacology of the Circulation		
October 1	Heart as a Pump	Dr. Stephenson
2	Frank-Starling Mech–Drugs Affecting Heart	Dr. Stephenson
	Anti-arrhythmic Drugs	Dr. Knuepfer
4	Systemic Circuitry and Hemodynamics	Dr. Ellsworth
5	Regulation of Arterial Blood Pressure	Drs. Macarthur/Panneton
8	Local Control of Perfusion	Dr. Ellsworth
9	Vascular Mediators	Dr. Macarthur
11	Platelets and Vascular Pathology	Dr. Stephenson
12	Homeostasis–Claude Bernard	Dr. Sprague
15	Hemorrhage	Dr. Sprague
17	Exam III	
Section #4: The Physiology and Pharmacology of the Lungs		
October 18	Roles of RBC & Hb in O_2 and CO_2 Transport and Leukocytes and Inflammation	Dr. Lechner
19	Lung Ventilation: Compliance & Resistance	Dr. Lechner
22	Lung Perfusion: Diffusion & Acid-Base	Dr. Lechner
23	Central & Peripheral Control of Ventilation	Dr. Panneton
25	Uniqueness of the lung	Dr. Sprague

TABLE 21 Advanced and Selected Topics in Pharmacological and Physiological Science: PPY511–514, *cont'd*

9:00-11:00 a.m. Room M360	Lecture Title	Faculty
Section #4: The Physiology and Pharmacology of the Lungs, cont'd		
26	Asthma	Dr. Sprague
29	Systems Integration: Exercise and Altitude	Dr. Lechner
30	Heart Failure	Dr. Sprague
November 1	Study Day	
2	Exam IV	
Section #5: The Physiology and Pharmacology of the Kidneys		
November 5	Renal Hemodynamids	Dr. Enkvetchakul
6	Glomerular Filtration	Dr. Enkvetchakul
8	Renal Tubular Function	Dr. Enkvetchakul
9	Hypertension	Dr. Westfall
12	Renal Mechanisms of Acid–Base Balance	Dr. Enkvetchakul
13	Cardio-renal interactions	Drs. Ellsworth, Stephenson, Sprague and Enkvetchakul
15	Study Day	
16	Exam V	
Section #6: The Physiology and Pharmacology of the Gastrointestinal System		
November 26	GI Hormones and Motility	Dr. Harkins
27	Gastric Acid, Secretion, Pancreas, and Bile	Dr. Harkins
29	Transport, Absorption, and Digestion	Dr. Harkins
30	Central and Peripheral Regulators of Energy Balance	Dr. Gamber
December 3	Lab/Demonstration	Dr. Harkins
4	Paper Discussion	Dr. Harkins
7	Exam VI	
14	Comprehensive Final	All Faculty
January 13	Make-Up Final (if required)	All Faculty

PPY513/514 taken from Spring 2001

2nd Semester	Title	Faculty
Section #7: The Physiology and Pharmacology of the Endocrine System and Cancer		
January 29	Pituitary	Dr. Haas
30	Endocrine Control of Metabolism	Dr. Samson
February 1	Drugs Used in Diabetes	Dr. Voigt
2	Adrenal	Dr. Chrivia
5	Thyroid Physiology/Endocrine Control of Male Reproduction	Dr. Haas
6	Spermatogenesis/Male Reproduction	Dr. Haas
8	Female Reproduction	Dr. White
9	Calcium Homeostasis	Dr. Harkins
12	Pharmacology of Reproduction	Dr. Chrivia
13	Student Presentations	Dr. Harkins
15	Lab/Demonstration	Dr. Harkins
16	Study Day	
19	Exam I	
20	Cancer I	Dr. Chrivia
22	Cancer II	Dr. Baldassare
23	Cancer III	Dr. Baldassare

TABLE 21 Advanced and Selected Topics in Pharmacological and Physiological Science: PPY511–514, *cont'd*

2nd Semester	Title	Faculty
26	Cancer IV	Dr. Chrivia
27	Cancer V	Dr. Chrivia
March 1	Exam II	
Section #8: The Physiology and Pharmacology of the Central Nervous System		
2	Neurodevelopment I	Dr. Voigt
5	Neurodevelopment II	Dr. Voigt
6	Receptors I	Dr. Voigt
8	Receptors II	Dr. Voigt
9	Receptors III	Dr. Voigt
12	Exam III	
13-16	Spring Break	
19	Overview of the Brain I	Dr. Zahm
20	Overview of the Brain II	Dr. Zahm
22	Sensory systems / physiology I	Dr. Ariel
23	Sensory systems / physiology II	Dr. Panneton
26	Sensorimotor systems	Dr. Ariel
27	Reward system	Dr. Zahm
29	Learning / memory	Dr. Ariel
30	Study Day	
April 2	Exam IV	
3	Overview of neuropharmacology	Dr. Westfall
5	Dopaminergic systems / Parkinsons	Dr. Neckameyer
9	Dopaminergic neuropharmacology	Dr. Neckameyer
10	Serotonergic systems / pharmacology	Dr. Voigt
12	Pain / opioids / analgesic	Dr. Voigt
13	GAGAergic systems / sedative–hypnotics	Dr. Voigt
16	Modification of excitability: anti-convulsants	Dr. Voigt
17	Study Day	
19	Exam V	
27	Comprehensive Final	All Faculty
May 29	Make-up for Comprehensive	All Faculty

TABLE 22 Syllabus for Human Physiology: PPY254 (Fall 2012)

Date	Day	Topic	Lecturer
Note: The assignments refer to the assigned textbook, *Human Physiology*, 12th ed., by Stuart Ira Fox.			
Aug 27	Mon	Homeostasis (pp. 2-10 and optional reading 10-21)	Baldassare
Aug 29	Wed	Chemical Composition of the Body (pp. 25-46)	Baldassare
Aug 31	Fri	Cell Structure (pp. 51-71)	Baldassare
Sept 3	Mon	Labor Day–no lecture	
Sept 5	Wed	Genetic Control (pp. 51-71)	Baldassare
Sept 7	Fri	Enzymes (pp. 88-96)	Baldassare
Sept 10	Mon	Metabolism–1 (pp. 96-123)	Chrivia
Sept 12	Wed	Metabolism–2 (pp. 96-123)	Chrivia
Sept 14	Fri	Transport Across Membranes (pp. 129-134, 140-146)	Baldassare
Sept 17	Mon	Osmosis/Osmotic Pressure (pp. 135-140)	Baldassare

TABLE 22 Syllabus for Human Physiology: PPY254 (Fall 2012), *cont'd*

Date	Day	Topic	Lecturer
Sept 19	Wed	Cellular Signaling (pp. 151-153, 318-326)	Baldassare
Sept 21	Fri	Review in normal classroom	
Sept 24	Mon	Exam 1 LRC A, B, C Section 1: 7:20-8:35 am Section 2: 1:50-3:05 pm	
Sept 26	Wed	Endocrine -1 (pp. 311-317, 327-343	Chivia
Sept 28	Fri	Endocrine -2 (pp. 311-317, 327-343)	Chrivia
Oct 1	Mon	Resting/Action Potentials–1 (pp. 146-150, 170-178)	Egan
Oct 3	Wed	Resting/Action Potentials–2 (pp. 146-150, 170-178)	Egan
Oct 5	Fri	Synapses (pp. 178-199)	Baldassare
Oct 8	Mon	CNS structure and neuron communication (pp. 160-169, 206-209, 222-235)	Westfall
Oct 10	Wed	Peripheral Nervous System & Sense Organs (pp. 240-258)	Westfall
Oct 12	Fri	Autonomic Nervous Sys & Adrenals (pp. 240-258)	Westfall
Oct 15	Mon	Review: 9-11 am lecture Hall 4 in SOM 1:30-3pm LRC A, B, C	
Oct 17	Wed	Exam 2 LRC A, B, C Section 1: 7:20-8:35 am Section 2: 1:50-3:05 pm	
Oct 19	Fri	Platelet Physiology & Coagulation (pp. 401-407, 410-414)	Baldassare
Oct 22		Fall Break–no lecture	
Oct 24	Wed	Contractile Properties of Muscle (pp. 355-373)	Baldassare
Oct 26	Fri	Striated, Smooth, & Cardiac Muscle (pp. 373-392)	Baldassare
Oct 29	Mon	Cardiovascular Physiology–1 (pp. 414-432, 437-439)	Baldassare
Oct 31	Wed	Cardiovascular Physiology–2 (pp. 445-459, 445-446)	Baldassare
Nov 2	Fri	Cardiovascular Physiology–3 (pp. 461-466, 469-480)	Baldassare
Nov 5	Mon	Pulmonary–1 (pp. 414-415, 525-533)	Lechner
Nov 7	Wed	Pulmonary–2 (pp. 530-539)	Lechner
Nov 9	Fri	Pulmonary–3 (pp. 539-546, 562-566)	Lechner
Nov 12	Mon	Review in normal classrooms	
Nov 14	Wed	Exam 3 LRC A, B, C Section 1: 7:20-8:35 am Section 2: 1:50-3:05 pm	
Nov 16	Fri	Renal Physiology–1 (pp. 574-588)	Enkvetchakul
Nov 19	Mon	Renal Physiology–2 (pp. 588-602)	Enkvetchakul
Nov 21-23	Wed, Fri	Thanksgiving break–no lectures	
Nov 26	Mon	Renal Physiology–3 (pp. 602-611)	Enkvetchakul
Nov 28	Wed	Female Reproductive Physiology (pp. 715-725)	Samson
Nov 30	Fri	Male Reproductive Physiology (pp. 705-715)	Samson
Dec 3	Mon	Gastrointestinal Physiology–1 (pp. 612-627)	Harkins
Dec 5	Wed	Gastrointestinal Physiology–2 (pp. 627-647)	Harkins
Dec 7	Fri	Physiology of Hearing (pp. 275-286)	Chrivia
Dec 8,9	Sat, Sun	Review 9-11 a.m. Lecture Hall 4 in SOM 2-4 pm LRC A, B, C	
Dec 10	Mon	Exam 4 LRC A, B, C Section 1: 7:20-8:35 am Section 2: 1:50-3:05 pm	

TABLE 23 Physician's Assistants Program, Clinical Pharmacology (1991–92)

Date	Day	Time	Topic	Lecturer
Aug. 29	Thurs	8:00-8:50	General Introduction to Pharmacology	T. C. Westfall
Aug. 29	Thurs	9:00-9:50	General Principles	T. C. Westfall
Sept. 5	Thurs	8:00-8:50	Mechanisms of Drug Action	T. C. Westfall
Sept. 5	Thurs	9:00-9:50	Autonomic, Somatic and Central Nervous System	T. C. Westfall
Sept. 12	Thurs	8:00-9:50	Adrenergic Agonists	M. McAuley
Sept. 19	Thurs	8:00-9:50	Cholinergic Agonists and Antagonists	M. McAuley
Sept. 24	Tues	2:30-4:30	Exam I	
Sept. 26	Thurs	8:00-8:50	Autacoids	D. B. Houston
Sept. 26	Thurs	9:00-9:50	Anti-Ulcer, GI Drugs	D. B. Houston
Oct. 3	Thurs	8:00-8:50	Anti-Hypertensives	S. Connaughton
Oct. 3.	Thurs	9:00-9:50	Diuretics	J. R. Pawloski
Oct. 10	Thurs	8:00-8:50	Cardiac Glycosides	J. R. Pawloski
Oct. 10	Thurs	9:00 9:50	Anti-Anginals Anti-Coagulants	A. Stephenson
Oct. 17	Thurs	8:00-8:50	Anti-Arrhythmics	M. M. Knuepfer
Oct. 17	Thurs	9:00-9:50	Anti-Lipids; Anti-Anemics	A. Stephenson
Oct. 21	Mon	2:00-4:00	Exam II	
Oct. 24	Thurs	8:00-9:50	Opioids/Drug Abuse	L. Holt
Oct. 29	Tues	3:00-3:50	Drug Abuse	L. Holt
Oct. 29	Tues	4:00-4:50	Sedative Hypnotics; Anti-Anxiety	D. Evans
Oct. 31	Thurs	8:00-9:50	Anti-Epileptic; Anti-Parkinson	L. McMahon
Nov. 7	Thurs	8:00-9:50	Anti-Psychotic; Anti-Depressant	D. Evans
Nov. 12	Tues	2:00-3:50	Anti-Microbial I	A. H. Gold
Nov. 14	Thurs	8:00-9:50	Anti-Microbial II	A. H. Gold
Nov. 18	Mon	2:00-4:00	Exam III	
Nov. 19	Tues	2:00-3:50	Anti-Neoplastics	M. C. Seinfeld
Nov. 21	Thurs	8:00-9:50	Glucocorticoids	Y. Kim
Nov. 26	Tues	2:00-3:50	Thyroid; Parathyroid; Insulin	A. H. Gold
Dec. 5	Thurs	8:00-9:50	Estrogens; Progestins	L. McMahon
Dec. 10	Tues	2:00-4:00	General Review	

TABLE 24 General Physiology: PPYG504 (Fall 2001)

Date	Topic	Faculty
Aug. 28	Membrane Transport	Dr. Ellsworth
30	Neurophysiology	Dr. Egan
Sept. 4	Neurophysiology	Dr. Egan
6	Hypothalamus & Special Senses	Dr. Forrester
11	Hypothalamus & Special Senses	Dr. Forrester
13	Blood and Immunity	Dr. Liu
18	Exam I	
18	Muscle	Dr. Ellsworth
20	Muscle	Dr. Ellsworth
20	Autonomic NS	Dr. Westfall
25	Autonomic NS	Dr. Westfall
25	Hemodynamics	Dr. Ellsworth
27	Peripheral Blood Flow I & II	Dr. Ellsworth
Oct. 2	Cardiovascular Overview	Dr. Forrester

TABLE 24 General Physiology: PPYG504 (Fall 2001), *cont'd*

Date	Topic	Faculty
2	Cardiac Cycle and Function	Dr. Forrester
4	Cardiac Cycle and Function	Dr. Forrester
4	Arterial Blood Pressure and Reflexes	Dr. Forrester
9	Arterial Blood Pressure and Reflexes	Dr. Forrester
9	Coronary Circulation	Dr. Forrester
11	Exam II	
11	Respiration	Dr. Lechner
16	Respiration	Dr. Lechner
18	Respiration	Dr. Lechner
23	Respiration	Dr. Lechner
25	Body Fluids	Dr. Liu
25	Kidney	Dr. Liu
30	Kidney	Dr. Liu
Nov. 1	Kidney	Dr. Liu
1	Acid Base	Dr. Liu
6	Temperature Regulation	Dr. Liu
6	Exercise	Dr. Forrester
8	Exam III	
8	Endocrinology	Dr. Chrivia
13	Endocrinology	Dr. Chrivia
15	Endocrinology	Dr. Gautam
20	Reproduction	Dr. M. Ruh
22	Thanksgiving–no lecture	
27	Reproduction	Dr. M. Ruh
27	Gastrointestinal	Dr. T. Ruh
29	Gastrointestinal	Dr. T. Ruh
Dec. 4	Gastrointestinal	Dr. T. Ruh
6	Gastrointestinal	Dr. T. Ruh
11	Exam IV	

TABLE 25 Advanced Pharmacology in Primary Health Care Nursing: NRN508

Session 1 9/7 General Principles
    a. Introduction; General Pharmacological Principles
    b. Drug Actions; Pharmacokinetics
    c. Adverse Drug Reactions

Session 2 9/14 Autonomic Drugs
    a. Introduction; Cholinergic Activators
    b. Cholinergic Blockers
    c. Adrenoceptor Stimulants; Blockers

Session 3 9/21 Cardiovascular Drugs
    a. Drugs for Hypertension
    b. Drugs for Vasodilation and Treatment of Angina
    c. Cardiac Glycosides and Treatment of Congestive Heart Failure

Session 4 9/28 Cardiovascular Drugs
    a. Antiarrhythmic Drugs
    b. Diuretics
    c. Review Session

TABLE 25 Advanced Pharmacology in Primary Health Care Nursing: NRN508, *cont'd*

Session 5	10/5	No Class
Session 6	10/12	Examination; Drugs Acting on Smooth Muscle
		a. Examination
		b. Histamine; Serotonin; Ergot Alkaloids
		c. Polypeptides; Prostaglandins; Bronchodilators and Agents to Treat Asthma
Session 7	10/19	Drugs Acting on the Central Nervous System
		a. Introduction; Sedative-hypnotic Drugs
		b. Alcohols; Antiepileptics
		c. General, Local Anesthetics
Session 8	10/26	Drugs Acting on the Central Nervous System
		a. Skeletal Muscle Relaxants; Therapy of Parkinson's Disease and other Movement Disorders
		b. Antipsychotic Drugs; Antidepressants
		c. Opioid Analgesics; Drugs of Abuse
Session 9	11/2	Drugs Acting on Blood; Anti-inflammatory Drugs
		a. Agents for Anemias
		b. Drugs for Coagulation Disorders; Hyperlipidemias
		c. Nonsteroidal Anti-inflammatory Drugs; Non-opioid Analgesics
Session 10	11/9	Endocrine Drugs
		a. Review session
		b. Hypothalamic and Pituitary Hormones
		c. Thyroid, Antithyroid Drugs
Session 11	11/16	Examination; Endocrine Drugs (cont'd)
		a. Examination
		b. Corticosteroids, Gonadal Hormones
		c. Pancreatic Hormones; Antidiabetic Agents
Session 12	11/23	Thanksgiving
Session 13	11/30	Chemotherapeutic Drugs
		a. Principles of Antimicrobial Action; Penicillins, Cephalosporins
		b. Aminoglycosides, Tetracyclines, Chloramphenicol, Polymyxin
		c. Antimycobacterial, Sulfonamides
Session 14	12/7	Chemotherapeutic Drugs (cont'd)
		a. Antifungal, Antiviral and Miscellaneous Agents
		b. Basic Principals of Antiparasitic Chemotherapy, Antiprotozoal Drugs
		c. Antihelmintic Drugs
Session 15	12/14	Chemotherapeutic Drugs (cont'd), Review Session
		a. Cancer Chemotherapy
		b. Immunopharmacology
		c. Review Session
Session 16	12/21	Final Exam

TABLE 26 Human Systems Physiology, Medical Anatomy & Physiology (Fall 2012)

Date	Day	Time	Topic	Lecturer
(Two-hour Lectures, usually on Tuesdays and Thursdays, three non-cumulative exams, and a final)				
Unit #1: Principles of General and Cellular Physiology				
8/28	Tues	9:00-9:50	Chap. 1: Principles of Homeostasis	Dr. Sprague
8/28	Tues	10:00-10:50	Chap. 2: Membrane Transport	Dr. Baldassare
8/30	Thur	9:00-10:50	Chap. 3: Fundamentals of Cell Signaling	Dr. Baldassare
9/4	Tues	9:00-10:50	Chap. 4: Membrane & Action Potentials	Dr. Egan
9/6	Thur	9:00-10:50	Chap. 5: Autonomic Nervous System I	Dr. Westfall
9/11	Tues	9:00-10:50	Chap. 6: Autonomic Nervous System II	Dr. Westfall
9/13	Thur	9:00-9:50	Chap. 7: Platelet Physiology	Dr. Stephenson
9/13	Thur	10:00-10:50	Chap. 8: Coagulation Pathways	Dr. Stephenson
9/18	Tues	9:00-10:50	Chap. 9: Building a Blood Vessel	Dr. Sprague
9/20	Thur	9:00-10:50	Chap. 10: Muscle Physiology	Dr. Ellsworth
9/25	Tues	9:00-10:50	Unit #1 Exam (16 hours of lecture)	Unit #1 Faculty
Unit #2: Cardiovascular and Respiratory Physiology				
9/27	Thur	9:00-10:50	Chap. 11: Cardiac Physiology	Dr. Stephenson
10/2	Tues	9:00-10:50	Chap. 12: Peripheral Circulation	Dr. Ellsworth
10/4	Thur	9:00-9:50	Chap. 13: Regulation of Arterial Blood Pressure	Dr. Macarthur
10/4	Thur	10:00-10:50	Chap. 14: Vascular Mediators	Dr. Macarthur
10/9	Tues	9:00-9:50	Chap. 15: Hypertension	Dr. Westfall
10/9	Tues	10:00-10:50	Chap. 16: Heart Failure	Dr. Sprague
10/11	Thur	9:00-9:50	Chap. 17: Erythrocyte Physiology & Hb Function	Dr. Lechner
10/11	Thur	10:00-10:50	Chap. 18: Inflammation & Immunity	Dr. Lechner
10/16	Tues	9:00-10:50	Chap. 19: Mechanics of Ventil.; Lung Compliance	Dr. Lechner
10/18	Thur	9:00-10:50	Chap. 20: Pulm. Circulation; Ventil./Perfus. Ratios	Dr. Lechner
10/23	Tues	9:00-10:50	Chap. 21: Alveolar Gas Exchange and Acid-Base	Dr. Lechner
10/25	Thur	9:00-10:50	Chap. 22: Ventil. Control in Rest, Exercise, & Sleep	Dr. Panneton
10/30	Tues	9:00-10:50	Unit #2 Exam (18 hours of lecture)	Unit #2 Faculty
Unit #3: Renal, Gastrointestinal, and Endocrine Physiology				
11/1	Thur	9:00-10:50	Chap. 23: Renal Physiology I	Dr. Enkvetchakul
11/6	Tues	9:00-10:50	Chap. 24: Renal Physiology II	Dr. Enkvetchakul
11/8	Thur	9:00-10:50	Chap. 25: Renal III: Body Fluids & Acid Base	Dr. Enkvetchakul
11/13	Tues	9:00-10:50	Chap. 26: Endocrine Physiology I	Dr. Chrivia
11/15	Thur	9:00-10:50	Chap. 27: Endocrine Physiology II	Dr. Chrivia
11/20	Tues	9:00-10:50	Chap. 28: Female Reproductive Physiology	Dr. Samson
11/27	Tues	9:00-10:50	Chap. 29: Male Reproductive Physiology	Dr. Samson
Unit #3: Renal, Gastrointestinal, and Endocrine Physiology, *cont'd*				
11/29	Thur	9:00-10:50	Chap. 30: Gastrointestinal Physiology I	Dr. Harkins
12/4	Tues	9:00-10:50	Chap. 31: Gastrointestinal Physiology II	Dr. Harkins
12/6	Thur	9:00-10:50	Make-up Lecture Date	
12/11	Tues	9:00-10:50	Unit #3 Exam (18 hours of lecture)	Unit #3 Faculty
12/14	Fri	9:00-12:00	Cumulative Course Final	Course Faculty

- **Barry M. Chapnick, PhD, University of Chicago**
  Cardiovascular pharmacology, peripheral vascular central mechanisms, eicosanoids, and endothelial biology.
- **Vincent A. Chiappinelli, PhD, University of Connecticut**
  Biochemical and cellular neuropharmacology, neuronal nicotinic acetylcholine receptor regulation and function, synaptogenesis, electrophysiology.
- **John Chrivia, PhD, University of Washington**
  Biochemical pharmacology and physiology, molecular biology and signal transcription factors, coactivators, and nuclear proteins.
- **Terrance M. Egan, PhD, Massachusetts Institute of Technology**
  Control of receptor and voltage-gated ion channels using electrophysiological and molecular biological approaches, and biology of purinergic receptors, especially P2X subtypes.
- **Mary L. Ellsworth, PhD, Albany Medical College**
  Microcirculatory hemodynamics and oxygen transport, cardiovascular science.
- **Decha Enkvetchakul, MD, University of Missouri School of Medicine**
  Renal physiology, structure and function of KATP channels, especially KirBac1.1.
- **Thomas Forrester, PhD, MD, University of Glasgow**
  Extracellular nucleotide release and actions, cardiovascular physiology.
- **Medha Gautam, PhD, Tata Institute of Fundamental Research**
  Cellular and molecular neurobiology; signaling interactions mediating synapse formation, development, and regulation.

TABLE 27 Representative Schedule and List of Topics for Drugs We Use and Abuse: BLA245

Session	Date	Lecture Topic	Session	Date	Lecture Topic
1	Aug 27	Orientation	25	Oct 22	Fall Break
2	Aug 29	History and Regulation	26	Oct 24	"Frozen Addict" Film
3	Aug 31	Nervous System	27	Oct 26	Opiates
4	Sept 3	Holiday	28	Oct 29	Hallucinogens I
5	Sept 5	Mechanisms of Drug Action	29	Oct 31	Hallucinogens II
6	Sept 7	Amphetamines	30	Nov 2	Drugs in Sports I
7	Sept 10	Cocaine	31	Nov 5	Drugs in Sports II
8	Sept 12	Depressants	32	Nov 7	Special Topics "Ecstasy"
9	Sept 14	Psycho-Therapeutics	33	Nov 9	Discussion (Groups E, F, G, H)
10	Sept 17	Special Topic–Migraine	34	Nov 12	Review Sessions for Exam 3
11	Sept 19	Review Session for Exam 1	35	Nov 14	Exam 3
12	Sept 21	Exam 1: Lectures 1-8	36	Nov 16	Drugs to Treat Infections HIV-AIDS
13	Sept 24	Alcohol I	37	Nov 19	Sexually Transmitted Diseases
14	Sept 26	Alcohol 2	38	Nov 21	NO CLASS
15	Sept 28	Marijuana	39	Nov 23	Thanksgiving
16	Oct 1	Inhalants	40	Nov 26	"Wonder Drugs"–Film
17	Oct 3	Tobacco	41	Nov 28	Antibiotics I
18	Oct 5	Caffeine	42	Nov 30	Antibiotics II
19	Oct 8	Over the Counter 1 (OTC)	43	Dec 3	Discussion 3 (Groups I, J, K, L)
20	Oct 10	Over the Counter 2 (OTC)	44	Dec 5	Discussion 4 (Groups M, N, O, P)
21	Oct 12	Special Topics: Asthma	45	Dec 7	Review Session for Exam 4
22	Oct 15	Discussion I (Groups A, B, C, D)	46	Dec 10	Exam 4: Lectures 25-30
23	Oct 17	Review Session for Exam 2	47	Dec 12	NO CLASS, Course Ends
24	Oct 19	Exam 2			

- **Karen Ann Gregerson, PhD, University of Nebraska School of Medicine**
  Cellular endocrinology, signal transduction in endocrine cells, regulable gene expression in the neuroendocrine system.
- **Alvin H. Gold, PhD, Saint Louis University**
  Endocrine science, cellular and molecular regulatory processes, carbohydrate metabolism.
- **Allyn Howlett, PhD, Rutgers University School of Medicine**
  Biochemical and cellular neuropharmacology, signal transduction mechanisms, cannabinoid receptor regulation and function.
- **Amy B. Harkins, PhD, University of Pennsylvania**
  Regulation of synaptic neurotransmissions, neuronal signaling and regeneration, cotransmission.
- **Yee S. Kim, PhD, Saint Louis University School of Medicine**
  Cellular control mechanisms, signal transduction in endocrine cells, vitamin D receptor regulation and function.
- **Mark Knuepfer, PhD, University of Iowa College of Medicine**
  Autonomic pharmacology and physiology, central cardiovascular regulation, electrophysiology, cardiovascular effects of cocaine and stress, mechanisms of hypertension.
- **Andrew John Lechner Jr., PhD, University of California, Riverside**
  Pulmonary physiology, acute lung injury, and the immunophysiology of sepsis.
- **Christian H. Lemon, PhD, SUNY Binghamton**
  Gustatory neural coding and sensory perception, neuroscience.
- **Maw-Shung Liu, PhD, University of Ottawa**
  Cardiovascular regulation, molecular and cellular mechanisms for cardiac and hepatic dysfunction in septic shock.
- **Andrew Lonigro, MD, Saint Louis University School of Medicine**
  Cardiovascular pharmacology and physiology; regulation and function of prostanoids and ecosinoids; pulmonary, coronary, and renal control mechanisms; and hypertension.
- **Heather Macarthur, PhD, William Harvey Research Institute, Saint Bartholomew's Medical College**
  Vascular control mechanisms, endothelial mediators, sympathetic neurotransmission, pathophysiology of Parkinson's disease.
- **Paul MacDonald, PhD, Vanderbilt University**
  Molecular endocrinology, vitamin D receptor and fat-soluble vitamin transcription and regulation.
- **Wendi Neckameyer, PhD, Rockefeller University**
  Drosophila molecular neurobiology, function and regulation of dopamine and serotonin.
- **Nicola C. Partridge, PhD, University of Western Australia**
  Cellular and molecular endocrinology, hormonal regulation of the osteoblast, role of metalloproteinases in bone.
- **William Michael Panneton, PhD, Ohio State University**
  Neuroanatomy and neurobiology of central autonomic control areas, physiology of the diving reflex, role of DOPAL in Parkinson's disease.
- **Mary Ruh, PhD, Marquette University**
  Chromatin receptor-binding factors and dioxin action, steroid hormone actions and regulation.
- **Thomas Ruh, PhD, Marquette University**
  Subcellular steroid hormone actions, estrogen/antiestrogen regulation and mechanism of action.
- **Daniela Salvemini, PhD, William Harvey Research Institute, St. Bartholomew's Medical College**
  Novel approaches for therapeutic development, pharmacology of reactive oxygen/nitrogen species, role of sphingolipids and adrenosinergic signaling in pain mechanisms.

- **Willis Kendrick (Rick) Samson, PhD, University of Texas Health Science Center**
  Neuroendocrine regulation of anterior pituitary hormone secretion, CNS regulation of fluid and electrolyte homeostasis, vasoactive peptides.

- **Leo C. Senay Jr., PhD, University of Iowa**
  Environmental physiology, fluid dynamics in exercise and heart stress, cardiovascular science.

- **Randy Stephen Sprague, MD, Saint Louis University School of Medicine**
  Pulmonary and cardiovascular physiology and pharmacology, role of the erythrocyte as a determinant of vascular resistance.

- **Alan H. Stephenson, PhD, Saint Louis University**
  Vascular control mechanisms, eicosanoids, pulmonary and cardiovascular pharmacology.

- **Willem van Beaumont, PhD, Indiana University**
  Environmental and exercise stress dynamics.

- **Mark Voigt, PhD, Saint Louis University**
  Biochemical and molecular neuropharmacology, molecular biology of serotonin and purinergic receptors, developmental neurobiology in zebrafish.

- **Thomas C. Westfall, PhD, West Virginia University**
  Biochemical neuropharmacology, monoamine and peptide regulation, autonomic and cardiovascular pharmacology, catecholamine and peptide roles in hypertension.

- **Daniel Scott Zahm, PhD, Milton S. Hershey Medical Center, Pennsylvania State University**
  Anatomy and function of the limbic system, extended amydala and nucleus accumbens, neurobiology of neurotensin, and substance abuse.

Research, Secondary, and Adjunct Faculty

- **William Banks**
  Neurobiology of the blood-brain barrier and aging, role of amaloid proteins in neurodegeneration, geropharmacology.

- **Paul Bajaj**
  Molecular biology of coagulation factors, hematology.

- **Lawrence Baudendistel**
  Local control of pulmonary vascular resistance, anesthesiology.

- **Jane Cox**
  Neurotransmitter receptors, molecular biology, and developmental regulation of glutamate and P2X receptors.

- **Sherin U. Devaskar**
  Molecular neuroendocrinology, regulation of neuronal insulin-like peptides, neuropeptides.

- **John Edwards**
  Renal physiology, ion channel regulation of kidney transport.

- **John Farah**
  Cellular and neuroendocrine science, cellular immunology.

- **Susan Farr**
  Behavioral pharmacology, mechanisms of neurodegeneration, aging and geropharmacology.

- **Steven Fleisler**
  Molecular biological and biochemical studies on the function and pathology of the retina and other eye diseases, lipid metabolism and retinal degeneration.

- **Michael Anne Gratton**
  Mechanisms controlling inner ear homeostasis.

- **Frank Hertelendy**
  Hormonal control of uterine and ovarian function, molecular and cellular endocrinology, prostanoid regulation and function.

- **Devin Houston**
  Biochemical pharmacology, cannabinoid receptor biochemistry and pharmacology.

- **Song-ping Han**
  Biochemical and cellular neuropharmacology, monoamine and peptide regulation and function.

- **Steven Herrmann**
  Cardiovascular pharmacology and physiology.

- **A. P. Li**
  Drug metabolism and toxicology.
- **John Koepke**
  Cardiovascular and neural pharmacology and physiology, neural control of renal function.
- **Linda Kusner**
  Mechanisms of myasthenia gravis, otolaryncology.
- **T. Lanthrone**
  Neuroendocrinology, electrophysiology.
- **David W. Moskowitz**
  Renal development and regulation, pathophysiological mechanisms of renal disease.
- **Ramakrishna Nemani**
  Molecular biology and genetics, biochemical endocrine science.
- **Felicia V. Nowak**
  Molecular endocrinology, hormonal and developmental regulation of novel reproductive neuropeptides.
- **Scott Martin**
  Use of microchip-based analytical devices to study monoamine and peptide regulation.
- **David Malone**
  Nephrology, bone metabolism, developmental biology.
- **Randy Strong**
  Molecular neurobiology and neuropharmacology, neurodegeneration disorders in aging.
- **Cheryl Quinn**
  Hormonal regulation of collagenase in the osteoblast.
- **L. James Willmore**
  Antiepileptic drug development and use, animal models of epilepsy.
- **Kung-Woo Peter Yoon**
  Neuropharmacology, electrophysiology, excitatory and inhibitory amino acid.

Extramural Support

Table 28 and Figure 28 show the total extramural support generated by full-time tenure-track faculty only in the Department of Pharmacological and Physiological Science for the period from 1991 to 2012. This was obtained from the dean's office and provided during the Annual Chair's Charter. Funding generated by research, secondary, or adjunct faculty is not included. Table 29 shows individual grants generated by tenure-track faculty for 1995–1996, and Table 30 shows the same information for 2000–01. These two years are representative and typical of the years from 1990 to 2013. During most of this time the department ranked in the top 50 percent of schools nationally. It reached its peak during the period from 1992 to 2000 and was ranked twenty-fifth in the nation four times, in 1995, 1996, 1998, and 2001. The peak funding amount of $4,793,394 was reached in 2004.

Formal Departmental Reviews

During the second part of the Westfall era (1990–2013), the Department of Pharmacological and Physiological Science underwent two formal reviews. The first covered the period from 1990 to 1995 and the second from 1997 to 2002.

Departmental Review, 1990–95

In preparation for the departmental review, the department carried out an extensive self-study and prepared three documents: 1) an academic plan, 2) a resource document, and 3) copies of the curriculum vitae of the primary faculty. The academic plan consisted of the following sections:

1. *The department*, consisting of a list of the faculty, a mission statement, history, a statement of strengths of the department as perceived by the faculty, weaknesses and areas needing improvement as perceived by the faculty, and finally a statement of the state of the department and a five-year plan
2. *MD instruction*, consisting of the medical physiology course and the medical pharmacology course
3. *Allied health professions, nursing and undergraduate education*
4. *Graduate PhD education*
5. *Faculty research*
6. *Faculty service activities*

TABLE 28 Extramural Support Generated by the Department of Pharmacological and Physiological Science (1991–2012)

Year	Extramural Support	NIH Rank in Pharmacology
1991	$ 2,212,398	NA
1992	$ 2,918,042	NA
1993	$ 2,621,892	35
1994	$ 2,938,562	26
1995	$ 3,480,766	25
1996	$ 3,197,753	25
1997	$ 3,872,382	30
1998	$ 4,166,273	25
1999	$ 3,844,492	30
2000	$ 4,120,869	37
2001	$ 3,714,748	25
2002	$ 4,166,970	41
2003	$ 4,157,686	56
2004	$ 4,793,394	66
2005	$ 4,185,415	65
2006	$ 3,719,218	77
2007	$ 3,361,472	74
2008	$ 3,192,807	63
2009	$ 3,782,495	63
2010	$ 3,395,146	79
2011	$ 3,000,542	85
2012	$ 3,002,148	88

(NA = Not Available)

7. *Individual faculty teaching activities*
8. *Faculty development*
9. *Finances of the department*
10. *Facilities*

The resource document (264 pages) consisted of similar sections but with much more detail. For instance, the MD instruction section included detailed lecture topics and schedules, policy statements, and individual faculty lecture topics. The graduate PhD education section listed graduates, students entering, admission and completion records, and complete profiles of all graduates and entering students. The faculty research section summarized research interests and contained a complete extramural funding record. The faculty service activities section summarized participation in university and national committees, editorial boards, study sections, invitations as speakers, and so forth.

Two committees were appointed by Dean Monteleone. *The Internal Committee*, also known as the Pharmacological and Physiological Science Review Committee (PPSRC), consisted of a chairperson, Dr. W. Michael Panneton of anatomy, Dr. Clifford Bellone of microbiology, Dr. Kyle E. Brown from pediatrics, Dr. Heinrich Joist from internal medicine,

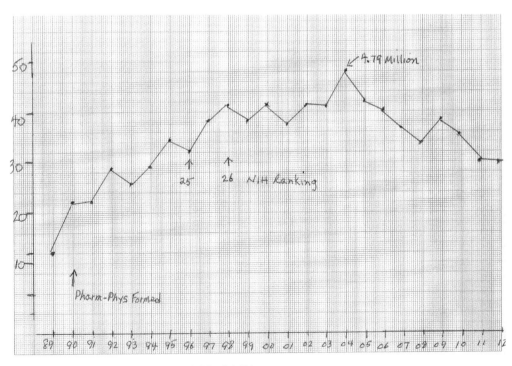

FIG. 28 Extramural funding (1989–2012)

Dr. George Vogler of pathology, and Gailya Barker from the administrative staff. *The External Committee* consisted of Dr. Harry Fozzard, professor and chair of the Department of Pharmacology and Physiology at the Pritzker School of Medicine, University of Chicago, and Dr. James E. Smith, professor and chair of the Department of Physiology and Pharmacology at the Bowman Gray School of Medicine, Wake Forest University.

The PPSRC met six times from October 1995 to March 1996. Before the first meeting, the committee received the resource documents prepared by the department. After an orientation meeting, the committee met by itself and with various members of the department. The chair of the committee, Dr. Panneton, selected input from the department and other departmental chairs, chose the External Committee from a list developed by Dr. Westfall and Dean Monteleone, and wrote a final report. This report was distributed to the external reviewers together with the resource documents prior to the site visit on April 25, 1996. During the site visit, the external reviewers met with members of the department, Dr. Westfall, departmental graduate students, selected medical students, the Internal Review Committee, and Dean Monteleone. Following the site visit, they prepared a written report. The reports of both the Internal Committee and the External Committee were given to Dr. Westfall, who shared them with the department. A follow-up written response and meeting with the Dean's Advisory Committee took place approximately one year later. The Executive Summary of the PPSRC and the Preamble of the External Review Committee's report are reproduced on pages 127–128.

Departmental Review, 1997–2002

The second formal departmental review took place in 2002 and covered the period from 1997 to 2002. As during the first review, both an Internal Committee and External Committee were formed. The Internal Committee, titled the Pharmacological and Physiological Science (PPS) Internal Analysis and Evaluation (A&E) Committee, consisted of chair Dr. G. Chinnedurai of molecular virology, Dr. John Tavis of molecular microbiology and immunology; Dr. Ratna Ray from pathology, Dr. Rita Heuertz, of internal medicine–pulmonology, Dr. Claudette Klein of biochemistry, Dr. Dan Tolbert of anatomy and neurobiology, and Gailya Barker from planning and operations.

The External Committee consisted of Dr. Eugene M. Silinsky, professor and chair of the Department of Molecular Pharmacology and Biological Chemistry at Northwestern University Medical School, and Dr. Steven P. Wilson, professor and chair of the Department of Pharmacology and Physiology at the University of South Carolina School of Medicine.

Just as with the first review, the department carried out an extensive self-study, which resulted in three sets of documents plus appendix material. The academic plan consisted of the following sections:

1. *The department*, which consisted of the mission statement, the history of the department, and strengths of the department (abstracted from faculty responses). These strengths included "leadership of the chairman," "collegiality among faculty," "quality of faculty and breadth and depth of expertise of the faculty," "a strong graduate program," "the seminar series, evening colloquia, and journal clubs," and "support staff." Weaknesses and areas needing improvement in the department, as abstracted from faculty responses, included "the graduate program," "teaching load not shared equally or variable," "diverse research interests," "location of departmental laboratories," and "infrastructure/institution-wide problems." Finally, this section included a five-year summary with the state of the department and a five-year plan.

2. *MD instruction*, which listed faculty participating in medical school teaching and described and discussed the following modules in which members of the department participate: cell biology, molecular biology and genetics, principles of pharmacology, cardiovascular system, nervous system, respiratory system, renal-urinary system, and endocrine/reproductive system.

Text continues on p. 130

TABLE 29 FY95 External Grants Report

Last Name	Grant #	Agency	Project Title	Begin Date	End Date	DC $	IC $	Total $
Andrade, R.	2R01 MH3985-07	NIMH	Mechanisms and Regulation of 5-HT Receptors in the CNS	09/01/94	08/31/95	$71,160	$32,734	$103,894
	1R01 MH49365-02	NIMH	Cholinergic Mechanisms and Receptors in Cerebral Cortex	07/01/94	06/30/95	$53,614	$24,662	$78,276
Subtotal						$124,774	$57,396	$182,170
Chapnick, B.	5R01 HL34036-09	NHLBI	Peripheral Vascular Control Mechanisms	07/01/94	05/31/95	$131,193	$60,349	$191,642
Subtotal						$131,193	$60,349	$191,642
Chiappinelli, V.	2R01 NS33135-01	NINDS	Electrophysiology of Presynaptic Nerve Terminals	07/01/94	06/30/95	$108,380	$46,928	$168,308
	2R01 NS17574-13	NINDS	Functional Properties of Neuronal Nicotinic Receptors	12/01/94	11/30/95	$159,844	$62,269	$222,113
		STRC	Neuronal Nicotinic Receptor Function after Embryonic Exposure to Nicotine	07/01/94	06/30/95	$79,318	$9,915	$89,233
Subtotal						$347,542	$19,112	$466,664
Chrivia, J.		PMAF	Characterization of the Physiological Role and Transactivational Properties of CBP, a Putative CREB Co-activator	01/01/95	12/31/95	$12,500	$0	$12,500
Subtotal						$12,500	$0	$12,500
Egan, T.		AHA-MO	Separation of the Metabolic Pathways Underlying Regulation by Cyclic-AMP of Chloride and Other Currents in the Heart	07/01/94	06/30/96	$30,000	$3,000	$33,000
Subtotal						$30,000	$3,000	$33,000
Ellsworth, M.	5R29 HL39226-07	NHLBI	Oxygen Affinity and Oxygen Transport in Striated Muscle	07/01/94	05/31/95	$76,003	$20,440	$96,443
	5R29 HL02602-03	NHLBI	Oxygen Transport and Hemodynamics in the Microvasculature	01/01/95	12/31/95	$60,150	$4,860	$66,610
Subtotal						$136,753	$26,300	$162,063

Last Name	Grant #	Agency	Project Title	Begin Date	End Date	DC $	IC $	Total $
Howlett, A.	2R01 DA06312-07	NIDA	Cannabinoid Receptor Ligands - SAR	08/01/94	07/31/95	$108,394	$49,861	$158,266
	5K05 DA00182-03	NIDA	Cannabinoid Receptor Pharmacology and Biochemistry	08/01/94	07/31/95	$91,125	$7,290	$98,416
	5R01 DA03690-11	NIDA	Cannabinoid Receptors in Neuronal Cells and Brain	09/01/94	07/31/95	$99,484	$45,763	$146,247
Subtotal						$299,003	$102,914	$401,917
Knuepfer, M.	5R01 DA05180-05	NIDA	Cardiovascular Effects of Cocaine	04/01/95	02/29/96	$123,673	$43,516	$167,189
Subtotal						$123,673	$43,516	$167,189
Lechner, A.		AHA-MO	Cytokine Modulation of Septic Shock and Acute Lung Injury during Immunosuppression	07/01/94	06/30/95	$24,964	$2,496	$27,460
Subtotal						$24,964	$2,496	$27,460
Lui, M.S.	2R01 HL30080-09	NHLBI	Substrate Metabolism in Isolated Myocytes in Shock	04/01/95	02/29/96	$110,389	$50,779	$161,168
Subtotal						$110,389	$50,779	$161,168
MacDonald, P.	7R29 DK49293-03	NIDDK	Coordinate Control of PTH Expression by Vitamins A & D	12/01/94	11/30/95	$70,000	$32,200	$102,200
Subtotal						$70,000	$32,200	$102,200
Neckameyer, W.		NARSAD	Regulation of Dopaminergic and Serotonergic Neurotransmission	07/01/94	06/30/96	$30,000	$0	$30,000
	IBN 9423616	NSF	Regulatory Mechanism of *Drosophila* Tyrosine Hydroxylase	04/01/95	03/31/96	$55,411	$24,277	$79,688
	IBN 9407599	NSF	Regulation of Neurotransmitter Transporters in *Drosophila*	09/01/94	08/31/96	$16,364	$1,638	$18,000
Subtotal						$101,775	$26,913	$127,688
Partridge, N.	1R01 DK47420-02	NIDDK	Nuclear Events in PTH Action on Bone Cells	01/10/95	12/31/95	$120,162	$54,366	$174,528
	NCC2-884	NASA	Effect of Microgravity on Bone Development	01/01/95	12/31/95	$92,410	$6,569	$98,979
	1R01 DK48109-02	NIDDK	Mechanisms of PTH Signal Transduction in Bone Cells	06/01/95	05/31/96	$107,164	$46,891	$154,055

TABLE 29 FY95 External Grants Report, *cont'd*

Last Name	Grant #	Agency	Project Title	Begin Date	End Date	DC $	IC $	Total $
	NAGW-4549	NASA	Skeletal Collagen Turnover by the Osteoblast	03/31/95	02/29/96	$92,826	$1,881	$104,487
	NGT-51145 Predoc Fellowship Sharon Bloch	NASA	Regulation of Collagenase Expression in Normal Differentiating Osteoblasts	09/01/94	08/31/95	$22,000	$0	$22,000
Subtotal						$434,562	$47,476	$582,038
Ruh, M.	1R01 ES05968-01	NIEHS	Receptor Binding Factor for the AH-Receptor	09/14/94	09/13/96	$80,000	$20,000	$100,000
Subtotal						$80,000	$20,000	$100,000
Stephenson, A.	1R01 HL52675-01	NHLBI	Cytochrome P-450 Arachidonate Metabolites in Lung Injury	04/01/95	02/29/96	$179,691	$4,470	$254,161
	2R01 HL32815-11	NHLBI	Lung Perfusion by Positron Emission Tomography	04/01/95	02/29/96	$9,170	$4,218	$13,388
Subtotal						$188,861	$78,688	$267,549
Voigt, M.		PMAF	Investigation of the Molecular and Cellular Biology of 5-Hydroxytryptamine 1 Receptors	01/01/95	12/31/95	$12,500	$0	$12,500
Subtotal						$12,500	$0	$12,500
Westfall, T.	ST32 GM08306-05	NIGMS	Pharmacological Sciences Research Training Grant	07/01/94	06/30/95	$56,377	$2,484	$58,861
	5R01 HL26319-14	NHLBI	Catecholamine Transmission in Vascular Smooth Muscle	12/01/94	11/30/95	$142,930	$65,748	$208,678
	5T32 NS07254-10	NINDS	Graduate Training in Neuropharmacology	07/01/94	06/30/96	$210,589	$15,010	$225,599
Subtotal						$409,896	$83,242	$493,138
TOTAL						$2,745,549	$711,262	$3,673,952

TABLE 30 FY2000 External Grants Report

Last Name	Grant #	Agency	Project Title	Begin Date	End Date	DC $	IC $	Total $
Baldassare, J.	IR01 GM59251-01 (D. Raben, PI)	Johns Hopkins Sub-Contract	Signaling Cascade of RhoA-Mediate PLD Activation	05/01/99	04/30/00	$45,584	$18,050	$63,634
Subtotal						$45,584	$18,050	$63,634
Chrivia, J.	5R01 DK52231-04	NIDDK	Activation and Repression of Transcription by CBP	02/01/00	01/31/01	$179,528	$83,656	$263,184
Subtotal						$179,528	$83,656	$263,184
Egan, T.	5R01 HLS 6236-03	NHLBI	Characterization of Cardiac Purinoceptors	07/01/99	06/30/00	$118,562	$56,910	$175,472
	0040199N	AHAEIG	Functional and Molecular Characterization of ATP-Gated Ion Channel-Receptors of Rat and Human Heart	01/01/00	12/31/00	$68,182	$6,818	$75,000
Subtotal						$186,744	$63,728	$250,472
Ellsworth, M.	5R01 HL56249-03	NHLBI	Erythrocyte-A Regulator of Microvascular Perfusion	12/01/99	11/30/00	$108,235	$51,953	$160,188
Subtotal						$108,235	$51,953	$160,188
Howlett, A.	5K0S DA00182-08	NIDA	Cannabinoid Receptor Pharmacology and Biochemistry	08/01/99	07/31/00	$91,125	$7,290	$98,415
	5R01 DA06312-11	NIDA	Cannabinoid Receptor Structure Activity Relationships	08/01/99	07/31/00	$186,933	$47,616	$234,549
	5R01 DA03690-15	NIDA	Cannabinoid Receptors in Neuronal Cells and Brain	06/01/99	03/31/00	$128,470	$59,671	$188,141
Subtotal						$406,528	$114,577	$521,105
Kim, Y.	5R01 DK49728-04 (K. Hruska, PI)	Wash. Univ. Consortium Agreement	Mechanism of Non-Genomic Action of Vitamin D	05/01/99	01/31/00	$75,891	$34,910	$110,801
Subtotal						$75,891	$34,910	$110,801

TABLE 30 FY2000 External Grants Report, *cont'd*

Last Name	Grant #	Agency	Project Title	Begin Date	End Date	DC $	IC $	Total $
Neckameyer, W.	IBN-9816566	NSF	Regulatory Mechanisms of *Drosophila* Tyrosine Hydroxylase	04/01/00	03/31/01	$43,932	$18,447	$62,379
	1R01 MH58724-01	NIMH	Molecular Analysis of the *Drosophila* GABA Transporters	08/01/99	07/31/01	$76,270	$36,610	$112,880
Subtotal						$120,202	$55,057	$175,259
Partridge, N.	5R01 DK47420-07	NIDDK	Nuclear Events in PTH Action on Bone Cells	01/01/00	12/31/00	$146,363	$70,254	$216,617
	5R01 DK48109-06	NIDDK	Mechanisms of PTH Signal Transduction in Bone Cells	06/01/99	05/31/00	$145,053	$68,857	$213,910
Subtotal						$291,416	$139,111	$430,527
Ruh, M.	5R01 ES05968-04	NIEHS	Receptor Binding Factor for the AH-Receptor	03/01/99	02/28/00	$137,364	$63,187	$200,551
Subtotal						$137,364	$63,187	$200,551
Ruh, T.	5R01 DK52478-02	NIDDK	Estrogen Receptor Interactions with Histones	09/01/99	08/31/00	$146,019	$70,089	$216,108
Subtotal						$146,019	$70,089	$216,108
Samson, W.	9950525N	AHA	Cardioprotectine Effects of Proadrenomedullin-Derived Peptides in Brain	09/01/99	12/31/00	$26,737	$2,674	$29,411
	RX-4235-662-SLU	Georgetown University Sub-Contract	GP120 Induced Neuroendocrine and Immune System Pathogenesis in HIV Disease	08/15/99	05/14/00	$16,000	$0	$16,000
Subtotal						$42,737	$2,674	$45,411
Stephenson, A.	5R01 HL32815-16 (D. Schuster, PI)	Wash. Univ. Consortium Agreement	Lung Perfusion by Positron Emission Tomography	03/01/00	02/28/01	$11,479	$5,510	$16,989

CHAPTER TEN

Last Name	Grant #	Agency	Project Title	Begin Date	End Date	DC $	IC $	Total $
	5R01 HL52675-06	NHLBI	Cytochrome P-450 Eicosanoids and Pulmonary Hemodynamics	03/01/00	02/28/01	$203,825	$97,836	$301,661
Subtotal						$215,304	$103,346	$318,650
Voigt, M.	2R01 N835534-04	NINDS	Molecular and Functional Analysis of ATP Receptor	04/01/00	03/31/01	$225,000	$108,000	$333,000
Subtotal						$225,000	$108,000	$333,000
Knuepfer, M.	2R01 DA05180-08	NIDA	Cardiovascular Effects of Cocaine	05/01/99	02/29/00	$120,552	$57,865	$178,417
Subtotal						$120,552	$57,865	$178,417
Liu, M.-S.	2R01 HL30080-12	NHLBI	Substrate Metabolism in Isolated Myocytes in Shock	07/01/99	06/30/00	$204,913	$65,564	$270,477
Subtotal						$204,913	$65,564	$270,477
Macarthur	IR0I HL61836-01	NHLBI	Deactivation of Catecholamines by NO, O_2 and Peroxynitrite	03/01/00	02/28/01	$125,000	$60,000	$185,000
Subtotal						$125,000	$60,000	$185,000
Westfall, T.	5R01 HL60260-02	NHLB	NPY induced Regulation of Sympathetic Neurotransmission	04/01/99	03/31/99	$210,256	$94,097	$304,353
	5T3 GM0830810	NIIGMS	Pharmacological Sciences Research Training Grant	07/01/99	06/30/00	$88,456	$5,276	$93,732
Subtotal						$298,712	$99,373	$398,085
FY00 TOTAL						$2,929,729	$1,919,140	$4,120869

The 1996 Formal Departmental Review

EXECUTIVE SUMMARY OF THE INTERNAL REVIEW COMMITTEE, APRIL 1996

The Department of Pharmacological and Physiological Science is a dynamic component of the Medical School of the St. Louis University. It honors its threefold mission of research, teaching, and service and should be lauded for its accomplishments in these areas. The faculty, staff, and students are led by an energetic and sincere chairman, Dr. Thomas Westfall, who serves as a role model for his departmental enterprise but who could possibly delegate more responsibility. The faculty share an almost universal pride in their department and find the atmosphere conducive to learning through the avenues of research and friendship.

The department is a relatively new entity, having been created a little more than five years ago with the merger of the independent Departments of Pharmacology and of Physiology. Most of the faculty have positive views of the merger although a vote on the Executive Committee has been lost with the elimination of a department.

Despite the current restrictive times for research funding, the faculty in the Department of Pharmacological and Physiological Science has been impressive with its success in garnering research dollars. Moreover, the department has been supportive of new faculty with seed monies and of more established faculty who have temporarily lost grant support with bridge funds. Such support has already paid off since some of the faculty have been refunded after losing research support for a brief hiatus. Also important in the research effort has been the collective cultivation of an environment where people ask scientific questions, are willing to explore to find the answers, and can share the excitement of each other's discoveries. It is anticipated that the research effort will grow as the younger cadre of newly-hired scientists reach their potential.

The department is responsible for the teaching of medical, graduate, and service courses in the undergraduate divisions. Feedback from students who have taken the courses has generally been favorable, and scores on national standardized tests for medical students have placed our students at or above the mean. The department has an active graduate program, granting approximately 4 PhD degrees a year. Most of these graduates have been successful in finding positions in academia and industry.

The Department of Pharmacological and Physiological Science provides its faculty facilities and space of excellent quality. Most faculty have ample research space in newly renovated rooms. Departmental finances are adequately documented and seem to be wisely spent for the general good of the department. However, the ebullient projections for future years may have to be tempered to parallel the harsh reality of present times.

The internal review committee thanks Dr. Westfall, the faculty, students, and staff of the Department of Pharmacology and Physiology for their tremendous cooperation during the review process.

PREAMBLE TO THE EXTERNAL REVIEWER'S REPORT, APRIL 1996

The external review committee met with graduate students, medical students, and faculty on April 25, 1996 to assess the functioning of the Department of Pharmacology and Physiology. The reviewers, Drs. Harry Fozzard and James Smith, found the department to have made excellent progress in the last five years. The accomplishments of the department in research, teaching, and service are outstanding. The faculty, students, and administration were very complimentary of the department and the chair. The integration of Pharmacology and Physiology has obviously been very successful as assessed by departmental performance and faculty and student satisfaction. The faculty and chair of the department are to be commended for their outstanding progress.

The external reviewers would also like to commend the PPSRC and its administration for the outstanding internal review after an extensive evaluation of the data base and extensive interviews with faculty and students. The departmental review process at SLU School of Medicine is an excellently organized and executed process. It is clear that these extensive reviews will continue to move the institution to higher levels of excellence in teaching, research, patient care, and other services.

The pharmacological & physiological science department was established in 1990 by merger of the two separate faculties under the chairmanship of Dr. Thomas Westfall. This new department has flourished, both in research scholarship and in teaching, creating a vital intellectual environment for the original faculty and recruiting six new young faculty. Under the talented leadership of Dr. Westfall it has grown into a fine department and is poised to achieve true excellence. We believe that it has the potential to join the top 20 Pharmacology and Physiology departments in the country in faculty quality and teaching excellence. Such achievement probably depends on availability of significant resources, which we will describe.

Our conclusion from analysis of the extensive materials made available to us and from meeting with the faculty and representative students is that this department and its leadership are healthy and are major contributors to university excellence. The Internal Review Committee has done a thorough job of summarizing the issues facing the department and we generally support their report. We will not repeat their descriptions, but will comment on their recommendations and add additional insights that we can provide from an external perspective.

3. *GEME students and undergraduate education*, which described General Physiology (PPYG504), Human Physiology (PPYS254), and Drugs We Use and Abuse (BLA245)
4. *Graduate PhD education*, which included a description of requirements for the PhD in pharmacological and physiological science, a description of the Graduate (Core) Program in Biomedical Sciences, a description of all the courses required or objectives, and the responsible conduct of research course. The section also listed all faculty and the courses they participated in. Finally, there was a summary of trainee credentials, applications enrollments, and graduates. (Details were in the resource document).
5. *Faculty research interests*, which listed extramural funding for each faculty member from 1998 to 2002, as well as scholarly interactions of the faculty within the department, the university, and the nation.
6. *Faculty service activities*
7. *Faculty teaching activities*
8. *Faculty development*
9. *Finances*
10. *Facilities*

The resource document (which went into great detail in its 264 pages) consisted of the following twelve sections: 1) A five-year summary; 2) organization and government; 3) MD instruction, which consisted of topics, scheduled faculty, and resources of the cell biology, principles of pharmacology, cardiovascular system, respiratory system, nervous system, renal and urinary system, and endocrine/reproductive system modules and electives offered by the PPS faculty; 4) GEME and undergraduate education consisting of information for General Physiology (PPY504), and Human Physiology (PPY254); 5) graduate PhD education; 6) faculty research interests, including a summary of research interests and all external grants and publications from 1997 to 2003; 7) faculty service activities, including departmental, university, and national committees; editorial boards; study sections and review groups; invitations to speak at national and international meetings and other institutions; and community service; 8) individual faculty teaching activities; 9) departmental space assignments; 10) personnel resources; 11) departmental finances; and 12) faculty impact.

Conduct of the Review

The Analysis and Evaluation Committee was appointed by Dean Patricia Monteleone in March 2002. The A&E Committee received the academic plan and resource document prepared by the department, the previous PPS Departmental Review from April 1996, and input from other departmental chairs. At the first meeting, the committee was given the charge of the dean, requested additional material, prepared questions to be asked of the chairman and faculty, and divided up subject areas of responsibility among the committee. The committee subsequently carried out individual interviews of the chair, faculty, graduate students, and medical students. Subcommittee summaries and recommendations were presented to the full A&E Committee, and a final report was prepared and submitted to the dean. The A&E Committee in consultation with the dean selected the External Committee.

The External Committee's site visit took place on June 20 and 21 of 2002. Committee members had been given copies of the resource material prepared by the department, the previous PPS Departmental Review from April 1996, and the report prepared by the Internal Review (A&E) Committee. The External Committee met with Dean Monteleone and held individual interviews with the department faculty, Dr. Westfall, and graduate and medical students. The following is the Executive Summary of the Internal and External Committee. The executive summaries of both the Internal and External Review Committees are reproduced on pages 130–132.

Seminars/Visiting Speakers Series (1979–2013)

One of the first things that Dr. Westfall established after becoming chair of the Department of Pharmacology in 1979 and later the Department of Pharmacological and Physiological Science in 1990 was to develop a vigorous and robust seminar, or visiting speakers, series. It was reasoned that this would greatly enhance the establishment of the department, hasten its recognition, and increase its overall quality. Inviting outstanding scientists and investigators to the department would result in two things: first, faculty, graduate students, and postdoctoral fellows would benefit from the expertise of the visitors as a result of individual meetings with the speakers and the formal presentations they delivered. Second, the visiting speakers would learn from the faculty, students, and fellows. It was apparent that these visiting scientists would ultimately be peer-reviewing their manuscripts and grant applications.

Throughout the Westfall era, approximately twenty to thirty outside speakers visited the department each year. A typical visit would last from one and a half to two and a half days, and the invited speaker would usually arrive in St. Louis in time for dinner on the first night with Dr. Westfall and several members of the faculty. The next day was spent with individual meetings between the speaker and faculty, having lunch with the graduate students, and giving their formal seminar (usually at 3:00 or 4:00 p.m.). The speakers would either leave that evening or have dinner with a different group of faculty and leave the following morning.

Between 1979 and 2014, more than one thousand distinguished speakers from the United States and abroad visited the department. Three groups deserve special mention: recipients of Nobel Prizes, speakers who were invited to give the D'Agrosa Lecture, and former graduates who were invited to give the Distinguished Alumni Lecture. Plaques were prepared with the speakers' names to commemorate the occasion and placed on display in M360, the large conference room.

Nobel Laureates

During the Westfall era, six speakers who were awarded the Nobel Prize in Physiology or Medicine or the Nobel Prize in Chemistry visited the campus: Sir John Vane, Ed Krebs, Al Gilman, Louis Ignarro, Ferid Murad, and Robert Lefkowitz. Table 31 lists each of these scientists, the year they were awarded the Nobel

The 2002 Formal Departmental Review

EXECUTIVE SUMMARY OF THE INTERNAL REVIEW COMMITTEE

The general impression of the Pharmacological & Physiological Science (PPS) Internal Analysis & Evaluation (A&E) Committee and that of a number of faculty members who responded to our request to comment on the Department is highly favorable. The consensus is that this is a well-run basic science department that has grown and thrived under the strong leadership of Dr. Thomas Westfall. This Department is an asset to the School of Medicine and to the University community in general. The A&E Committee is impressed with Dr. Westfall's dedication to the success of his faculty and students. He has a management style of leading by example and of being a "super-facilitator." He is an excellent model to his faculty and students. Dr. Westfall has an active hands-on role in running the Department and serves on many Departmental committees.

The performance of the Department in its three missions—research, teaching, and service—is strong. Research in the Department is diverse and well-funded. In the last five years the faculty has published 335 papers and generated over $20 million in external support. Research covers a wide range of topics in the fields of Pharmacology and Physiology, with a general theme of intercellular signaling and specific concentrations in Neuroscience, Endocrinology, and Cardiovascular Science. The Chairman also envisions future expansion in Cancer, one of the School of Medicine's focused areas for research.

Nationally, the Department ranks in the middle of similar departments in receipt of grants and contracts. Dr. Westfall acknowledged a small decline in this ranking due to the departure of a few senior faculty members but expressed confidence that the ranking will improve in the coming years. The A&E Committee noted that a few senior faculty members, who are not PIs in extramural grants, are very active co-investigators in funded research projects, and recover substantial portions of their salary from extramural grants. The Committee is satisfied with the mentoring efforts of the Department to see that every faculty member is externally funded.

Education of students, mainly at the medical and graduate levels, is a major endeavor for this Department. A few general concerns regarding MD and graduate education were noted. Faculty from this Department teach in most Phase I/II School of Medicine courses and modules, direct four modules, and are generally acknowledged by the students for their dedication and knowledge base. It is evident that the faculty strives to provide state-of-the-art medical education in the pharmacological and physiological sciences. The faculty did express grave concerns about the recent performance of our medical students on the USMLE I exam, the lack of fundamental physiological and pharmacological concepts being taught in some modules, and the paucity of feedback about faculty from most course/module directors to the Chairman regarding their teaching.

The Department is well known for its congenial and nurturing attitude toward its students and has been highly successful in attracting graduate students through the Core Graduate Program despite a shrinking applicant pool. Both the faculty and students acknowledged major difficulties with the Core Graduate Program and expressed the need for these to be formally addressed as soon as possible. The A&E Committee supports this proposal. The faculty and students also universally agreed that once a student has entered the Department, the student can obtain a top-notch education, but their overall experience suffers from a variable standard of rigor in the courses and in the preliminary examinations. It is also noted that teaching responsibilities are not evenly distributed among the faculty. There is a general opinion that the students are well trained for life in academics after graduation but may be ill prepared for private industrial positions.

The faculty is exemplary in providing scientific services within the University and service to the community. Interdepartmental relationships and external collaborations with other academicians are good. Several clinical faculty members have expressed gratitude for the faculty of this Department for their contributions to their research programs. It was noted that here is a lack of emphasis on technology transfer, little emphasis is given to entrepreneurism at the faculty level. Indeed, several faculty members were unable to define how to proceed with their own potential intellectual property ideas.

The Department derives its financial support from hard dollars, the Health Sciences Center (RSC) endowment, a second endowment from the William Beaumont Chair, and extramural grants to the faculty members. Most of the hard dollars and the HSC endowment incomes are used towards faculty salaries. The Beaumont endowment is used to support the salary of the Chairman. With a sustained level of University and extramural support, it should be possible to meet future financial needs adequately. The Chairman recognizes and appreciates the Dean's substantial support to his Department by means of attractive start-up packages for the recruitment of new faculty members. Faculty who temporarily lose grant support are eligible for competitive bridge funds available from the Dean. With this interim assistance, PPS faculty members have proven to be very successful in reacquiring extramural grant support. The Committee feels that the University policy of "lapsed" funds (i.e., return of unused salary back to the University) is a serious impediment to maintaining an emergency fund for support of needy faculty during difficult times. Further, this policy appears to diminish faculty enthusiasm in recovering additional salary.

The faculty is generally happy with their space allocation and research facilities. Although the facilities are currently adequate, they are aging and offer no room for expansion. Departmental space is problematic at times due to the "sprawl" of the Department, which is spread over three floors throughout the School of Medicine. The Department would like contiguous space. It is recommended that the deficiencies in the existing facilities be remedied and that the research space available to the Department be expanded consistent with the availability of new space through construction of the new research building.

Finally, the internal A&E Committee would like to thank Dr. Westfall, along with his faculty and staff, for their cooperation during this review process.

EXECUTIVE SUMMARY OF THE EXTERNAL REVIEW COMMITTEE, JUNE 2002

It is our impression that Saint Louis University (SLU) School of Medicine is very fortunate to have a department that maintains the consistently high quality of research and teaching as does the Department of Pharmacological and Physiological Science (PPS). Much of the credit for the development of this Department and the positive tone that permeates PPS goes to Dr. Thomas Westfall, an exceptional chairperson who leads by example. Dr. Westfall has created a collegial and productive community of scholars in his Department. We thus fully agree with the conclusions in the internal report that PPS is an "asset to the School of Medicine and the University community in general." Indeed Dr. Westfall is a rarity; a nurturing chair whose goal is to remove obstacles from his faculty so they can flourish as independent scholar/scientists. Many of the salient issues concerning the Department were presented in the very thorough review of PPS conducted by the Internal Analysis & Evaluation Committee. Our report will focus on observations and recommendations that we feel need to be emphasized or elaborated upon further.

Discussions with the Chairperson:

In view of the collegiality of the Department, it is essential to keep all of the active scholars in PPS together when the new research building is opened. The productive faculty in PPS should either be moved to contiguous space in the new building, or be allowed to occupy contiguous space in the current building.

Additional incentives are imperative to encourage faculty to pay salary from their research grants and to help grow the enterprise.

report. Specific suggestions are provided later in this report.

With respect to the integrated medical curriculum, more control of the curriculum needs to be under the aegis of the basic science chairs. The module directors need to provide the Chair with critiques of the lectures presented by the PPS faculty.

A transgenic facility needs to be provided and supported by University funds.

Dr. Westfall (and SLU as a whole) should not be required to use Water Tower Inn to accommodate potential faculty candidates (or potential Chair candidates). Such restrictions are a detriment to the SLU Medical Center—see discussions in report.

Departmental Resources and Faculty issues:
Dr. Westfall should delegate responsibilities and trust other faculty to assume greater positions of leadership in the SLU community. A formal mentoring process for newly recruited faculty should be instituted at the beginning of their tenure clock at SLU. A thorough review of the position of the Director of the Core Graduate Program should be conducted.

Research:
The Department faculty has been successful in obtaining national funding for research programs, in publishing the results of their work, and in receiving other indications of success in scientific research. The primary means of achieving funding has been individual investigator awards. To enhance funding and recognition, the Department and School should consider ways to develop collaborative research projects. The Department and School should also establish a mechanism of faculty incentives for research productivity.

Medical Education:
With the initiation of the new curriculum, control of the pharmacology and physiology curricular content and teaching by Department faculty has been largely removed from the control of the Department and its Chair. Although Department faculty members teach in the curriculum, the means to assess their performance appears to be lacking. As a consequence of these facts, data are largely lacking for assessing the outcome of the Department's medical teaching.

Graduate Education:
The Department has created an excellent environment for the training of graduate students. Although the Department needs to assess its second-year curriculum, both students and faculty members believe that Department training of graduate students is very positive. Frustrations with the graduate program mostly concern the lack of adequate student numbers and the director of the Core Graduate Program.

Prize, and the area of expertise for which the prize was awarded.

Louis D'Agrosa Memorial Lecture

This endowed lectureship is one of the most important of all the lectureships in the department. It is named in honor of a very popular professor, Dr. D'Agrosa, who was a member of the Department of Physiology at Saint Louis University School of Medicine from 1962, when he earned his doctorate, until his death in 1983. He studied the control of abnormal cardiac rhythms and was a pioneer in the use of bats to study microcirculation. Dr. D'Agrosa was also a popular and respected teacher and taught medical, dental, nursing, and allied health students. Members of the entire Saint Louis University community contributed to the endowment in his memory. The Louis D'Agrosa Memorial Lecture initially started in the Department of Physiology and was continued after the formation of the Department of Pharmacological and Physiological Science. Through the years, it became the hallmark lectureship sponsored by the

TABLE 31 Nobel Laureates Who Gave Seminars in the Department (1979–2014)

Name	Year Awarded the Nobel Prize	Affiliation at the Time	Area of Expertise
Sir John Vane	1982	Director, William Harvey Institute, St. Bartholomew's School of Medicine, University of London	Discoveries concerning prostaglandins and related biologically active substances
Ed Krebs, PhD	1992	Chair, Dept. of Pharmacology, University of Washington School of Medicine	Reversible protein phosphorylation as a biological regulatory system
Al G. Gilman, MD, PhD	1994	Chair, Dept. of Pharmacology, University of Texas, Southwestern Medical Center	Discovery of G proteins and their role in signal transduction
Louis Ignarro, PhD	1998	Professor, Dept. of Pharmacology, UCLA School of Medicine	Nitric oxide as a signaling molecule in the cardiovascular system
Ferid Murad, MD, PhD	1998	Chair, Dept. of Pharmacology, University of Texas Health Science Center	Nitric oxide as a signaling molecule in the cardiovascular system
Robert Lefkowitz, MD	2012	Professor, Dept. of Biochemistry and Medicine, Duke University	G protein-coupled receptors

Department and Lectureship Fund. In addition to attracting outstanding speakers, it was accompanied by a lavish hosted reception and get-together, during which members of the department and Saint Louis University could interact with the speakers. Table 32 gives the name, affiliation, and title of many of the D'Agrosa lecturers through the years.

Distinguished Alumni Lectureship

A third lectureship of note was established to honor former PhD graduates who had distinguished themselves in their careers. During their visit they were awarded a plaque to mark the occasion at a reception following their seminar. To date, five graduates have been honored. Their name is followed by the year and discipline of their degree and title when they gave their lecture.

- Dr. Stuart Dryer, PhD, 1985 (Pharmacology); John and Rebecca Moore's Professor and Chair, Department of Biology and Biochemistry, University of Houston
- Dr. Jason Weber, PhD, 1997 (CMB); Professor, Division of Oncology, Department of Internal Medicine, Washington University School of Medicine
- Dr. Deborah Anne Roess, PhD, 1982 (Physiology); Professor, Department of Physiology and Endocrinology, Colorado State University
- Dr. Gonzalo E. Torres, PhD, 1999 (Pharm Phys Sci); Professor, Department of Pharmacology, University of Florida School of Medicine
- Dr. Lori (McMahon) Wakefield, PhD, 1993 (Pharm Phys Sci); James E. Lowder Professor of Neuroscience; Dean, Graduate School, University of Alabama-Birmingham

Faculty Meetings and Retreats

In the early years of the Westfall era, faculty meetings were not held frequently. The size of the faculty was small, labs were in close proximity to each other, and faculty interacted often. There were also ample opportunities for information flow to take place at seminars, journal clubs, B&B sessions, and medical and graduate student course meetings. Faculty meetings were held every three months or so, or whenever needed. As the department grew, especially after the formation of the Department of Pharmacological and Physiological Science in the 1990s, it was decided that frequent and regular faculty meetings would be desirable. Faculty began to meet on a monthly basis,

TABLE 32 D'Agrosa Lectures

Walter C. Randall
Loyola University
Autonomic Systems of Cardiovascular Regulation

John T. Shepard
Mayo Clinic
Coronary Vasospasm—A Scientist's Viewpoint

Robert J. Lefkowitz
Duke University
Receptors and Rhodopsin:
Shedding New Light on an Old Subject

Urs S. Rutishauser
Case Western Reserve
N-CAM as a Regulator of Cell-Cell Interaction

Anthony R. Means
Baylor College of Medicine
Mechanisms and Consequences of Calmodulin as a Transducer of the Calcium Signal

Norman C. Staub
University of California, San Francisco Pathophysiology of Pleural Effusions

Marc E. Lippman
Georgetown University
Growth Regulation of Malignant Mammary Epithelium

Alfred G. Gilman
University of Texas, Dallas
G Protein and Regulation of Adenylyl Cyclase

Thomas Curran
Roche Institute
Oncogenes, Transcription, Regulation and the Brain

Lee Limbird
Vanderbilt University
Achieving Specificity in Signal Transduction

Heidi Hamm
Northwestern University
Structure and Function of G Proteins and Role in Signal Transduction

Kenneth Korach
National Institute of Environmental Health Sciences
Understanding Estrogen Hormone Action and Receptor Mechanisms in Estrogen Receptor Knock-Out Mice

Geoffrey Burnstock
Royal Free & University College
Medical School of University College, London
Purinergic Signalling

Kevin J. Catt
National Institute of Child Health and Human Development of Angiotensin II Receptors: The Odd Couple of the GPCR Superfamily

John D. Scott
Vollum Institute
Oregon Health Sciences University
The Molecular Architecture of Kinase Phosphatase Signaling Complexes

W. Jonathan Lederer
University of Maryland Biotechnology Institute
When Sparks Fly—Calcium Signaling In the Heart

David J. Mangelsdorf
University of Texas Southwestern Medical Center
The Role of Nuclear Receptors and Lipid Metabolism

Hershel Raff
Medical College of Wisconsin
The Endocrinology of Stress in the Neonate

Benita Katzenellenbogen
University of Illinois Urbana
Estrogens and Estrogen Receptor Actions in Human Health

William C. Sessa
Boyer Center for Molecular Medicine
Yale University School of Medicine
Insights into Control of Blood Flow and Angiogenesis

Aaron J. W. Hsueh
Stanford University School of Medicine
Hormone-Receptor Studies in the Postgenomic Era

Joseph T. Bass
Department of Medicine
Feinberg School of Medicine
Northwestern University
Circadian Integration of Energetics and Metabolism

Carol Elias
Department of Molecular and Integrative Physiology
University of Michigan School of Medicine
Nutrition and Growth: Search for Key Hypothalamic Pathways in Pubertal Development

Susan G. Amara
University of Pittsburgh
A New Take on Uptake: Neurotransmitter Transporters and the Activation of Intracellular Signaling Pathways by Amphetamine

and occasionally more frequently, to make sure that all faculty were kept abreast of departmental, school, university, and national activities and that all faculty felt part of the governance of the department. Meetings usually took place for one and a half to two hours. Agendas were prepared and there was ample opportunity for discussion of pertinent issues and decision-making. All faculty (tenure, research, and secondary track) were invited to the meetings.

Table 33 shows a representative agenda for a faculty meeting during the Westfall era. This gives a good idea of what issues were discussed, what reports were made, and how the meetings were conducted over the years.

In addition to faculty meetings, occasional retreats were also held off campus. During the Westfall era, five formal retreats were held. The first involved the Department of Pharmacology and the other four took place after the formation of the Department of Pharmacological and Physiological Science. Table 34 shows the dates and locations of these retreats. The purpose of the retreats was to allow ample time for all faculty to discuss and have impact on all departmental activities, functions, and decisions. Graduate students and postdoctoral fellows also had the opportunity to give poster presentations and talks. Ample time was available for social interactions among faculty, students, postdocs, and staff.

Social Activities

At Dr. Westfall's previous positions at West Virginia University, the Collège de France, and the University of Virginia, social interactions were very important and greatly added to the atmosphere of these departments. Developing a similar atmosphere in the department at Saint Louis University was a high priority. Numerous interactions were developed in the

TABLE 33 Agenda of the Department of Pharmacological and Physiological Science Faculty Meeting, January 23, 2003

- Announcements
  - AOA Forum; February 10; 1-4 p.m.; Lecture Room A
  - Graduate Student Association Research Symposium March 21; Abstracts Due February 3
  - Faculty Assembly; January 28; 3:30 p.m.; Anheuser Busch Auditorium; Cook Hall
  - New Chairs:
    - Dr. Mark Varvanes; Otolaryngology
    - Dr. Mark Comunale; Anesthesiology
    - Dr. Burton Moed; Orthopedics
  - Ongoing Searches
    - Anatomy and Neurobiology (Dr. Jack Selhorst, Chair)
    - Radiation Oncology (Dr. Jim Hardin, Chair)
  - SLU - Number 1 in number of doctorates offered (July 1, 2000-June 31, 2001): 146
- Research Building Update (Price Waterhouse Coopers - Business Plan)
- CMC Business
  - Joint Committee of Phases
  - Grading Policy
  - Proposed change in Chair of Committee (Discussed in March; vote in June)
- International Association of Medical Science Educators (IAMSE)
- Next meeting; July 19-22, 2003; Georgetown University
- Call for Topics; Meeting- July 10-13, 2004; Tulane University
- Letter to Bill Mootz concerning Departmental Teaching in Medical Curriculum Physiology in the Medical School Curriculum
- Plans for Review - Core Graduate Program
- Requests:
  - Director of BBSG592 (Colloquia)
  - Papers for BBSG504 (Special Topics)
- Doctoral Dissertation Fellowships
- Microarray Facility
- Emergency Contact Information (Please respond to MAK)
- Scholarly Activities Database
- Shared Instrument Grant Proposal?
- Conflict of Interest Policy and Disclosure
  - Faculty Manual Amendments
- Recruitments
- D'Agrosa Lecture
- Dean's December Items
- Old Business
- New Business

department that helped to create the warm, family atmosphere that became a tradition and hallmark of the Department of Pharmacology and later the Department of Pharmacological and Physiological Science. The following is a brief description of some of the more important social functions and traditions.

Welcome Party for New Staff, Faculty, Students, and Fellows

In the early fall a party was held to welcome all newcomers to the department. This included new staff, faculty, graduate students, postdoctoral fellows, and visiting professors. Each person was introduced by Dr. Westfall. Snacks and drinks (wine, beer, soft drinks, etc.) were provided by the department, and the sessions were held in the conference room. This became a yearly tradition and helped get each year started.

PhD Graduation Celebration and Awarding of a Saint Louis University Chair

At the conclusion of each PhD student's seminar and dissertation defense, a party of celebration was held jointly sponsored by the mentor and the department. The highlight of the celebration was the awarding to each student of a captain's chair or rocking chair that had the crest of Saint Louis University, compliments of the department. This became one of the most cherished traditions generated in the department. It was a tradition carried out in the Department of Pharmacology at the University of Virginia and brought to the Department of Pharmacology and subsequently the Department of Pharmacological and Physiological Science by Dr. Westfall when he came to Saint Louis University as chair. This tradition is very much appreciated by each graduating PhD student, and there have been many testimonials by previous students of how special this was to them. As of this writing, 130 chairs have been awarded.

Annual Christmas Party and Celebration

Throughout Dr. Westfall's chairmanship, it was traditional during the Christmas season to decorate the hallways, bulletin boards, and doors of the third floor around the administrative area and conference room. In addition, a Christmas party was also held for all departmental members and friends. The main entrées (ham, turkey, etc.) plus beverages (soft drinks,

TABLE 34 Department Retreats: Dates and Locations

April 26-27, 1989	Double Tree Conference Center Chesterfield, MO
Jan. 3-4, 1996	Double Tree Conference Center Chesterfield, MO
Nov 21-22, 2002	Cedar Creek Conference Center New Haven, MO (Agenda and Schedules Included)
Nov 22, 2005	Manresa Center Saint Louis University Conference Center
Nov 20, 2007	Manresa Center Saint Louis University Conference Center

wine, and beer) were provided by the department, but it became a tradition that each person attending would bring a side dish or dessert. Having a department with national and international representation resulted in quite a variety of exquisite foreign dishes added to the bill of fare. It was also a tradition that in addition to members of the department, members of other departments, as well as housekeeping and the maintenance departments who do so much to keep the department running smoothly, attended. This party became a much-anticipated annual event. In the early days, the party was held in one of the rooms in the LRC, but later it was held in the "Big Conference Room," M360. For quite a few years all the basic science departments held such parties, and there was even some competition among the departments as to who could throw the best party. Unofficially, it appeared that the Department of Pharmacology and later the Department of Pharmacological and Physiological Science was always the big winner. Although the party in the department continued, other departments did not continue the tradition. One thing that helped the party was the fact that, as mentioned, there were many Christmas decorations all along the hallways surrounding the conference room, including colored lights along the ceilings, all of which added to the atmosphere.

Agenda of the Departmental Retreat
Cedar Creek Conference Center, New Haven, Missouri, November 21-22, 2002

Thursday, Nov 21, 2002
Session I: The Pines Conference Room:

9:00-9:20
- Summary of Departmental Review and Response by TC

9:20-10:00 **Topic: Medical School Teaching**
- Summary of Departmental Participation
- Pharmacology and Physiological Components (Core Contents of Individual Modules)

First Year Courses
- Cell Biology
- Genetics and Molecular Biology
- Principles of Pharmacology

Second Year Modules
- Nervous System
- Hematology
- Cardiovascular
- Respiratory
- Renal
- Endocrine/Reproductive
- GI
- Skin/Connective Tissue
- Intro to Medicine

10:00-10:45
- Are There Deficiencies?
- Are We Satisfied?
- Development of Subcommittees to Examine
- Proposal for Peer Review of Departmental Teaching
- Teaching Allocation of Resource Units to Department
- Electives

10:45-11:00 Morning Break

Session II: The Pines Conference Room (11:00-12:30)
Topic: Research
- History of Support
- How can we Increase Our Extramural Support? R0I; P0I; Foundation; Industry; Technology Transfer
- Resources:
  - What do We Need?
  - What are the Deficiencies?
  - How Can We Obtain Them?
- Incentive Compensation Plan
- Mentorship and Faculty Development

Session III - Pines Conference Room (1:30-3:00)
Topic: Departmental Graduate Program
- Prelim Exam
  - Current Rules and Model
  - What are the Problems/Concerns?
  - How Can We Improve the Exercise?
  - Timing
- Monitoring of Students and Thesis Committees
  - Proposal by M. Ruh
- Second Year Course
  - Participation by Faculty
  - Objectives?
  - Are Topics Being Adequately Covered?
  - What Topics should be Expanded/Deleted?
- Direct Recruiting into the Department
  - Should We Do It?
  - How Should We Do It?

Session IV - Pines Conference Room (3:15-5:00)
Topic: Core Graduate Program
- Summary of Core Curriculum
- Summary of Departmental Participation
- Input by the Department into What is Being Taught, Especially Integrative Physiology/Pharmacology Section
- Strategy to Preserve What is in the Core Program that Impacts the Department?
- Rotations
- Recruiting

Friday, Nov. 22, 2002
Session V - Pines Conference Room (9:00-10:30)
Topic: Student Reports
- Second Year Course
- Prelim Exam
- Core Graduate Program Courses
- Report on BLA245 "Drugs We Use and Abuse"
- Other Items?

Additional Item
- What More Should Be on the Web Page?

Student Posters
TITLES AND ROOM ASSIGNMENT

1. NPY Modulation of Catecholamine Synthesis and newly Synthesized Release - Adepero Adewale (Advisor: T. C. Westfall) (Hawthorne)
2. Bovine Retinal Pericytes Undergo Apoptosis in Response to Elevated Concentrations of Insulin - Melanie Berken (Advisor: Andrew Lonigro) (Locust)
3. Effects of Chronic Cocaine Treatment on Hemodynamic Responses to Acute Cocaine and Lipopolysaccharide - Tracy Bloodgood (Advisor: Mark Knuepfer) (Hawthorne)
4. Regulatory Mechanisms of *Drosophila* Tryptophan-Phenylalanine Hydroxylase - Chandra Coleman (Advisor: Wendi Neckameyer) (Willows)
5. Role of Rapsyn Domains in its Association with Synaptic Molecules - Sallee Eckler (Advisor: Medha Gautam) (Hawthorne)
6. Cannabinoids and the Modulation of Neuropeptide Y Release from the Hypothalamus - Kevin Gamber (Advisor: T. C. Westfall) (Locust)
7. cc-Thrombin Mediates the Activation of the Ras/Erk Cascade in Chinese Hamster Embryonic Fibroblasts (IIC9) Cells by Release of Selective Gpy Subunits from Gi2 - Reema Goel (Advisor: Joe Baldassare) (Willows)
8. Keratinocyte Interleukin-IP Expression Is Mediated by the Aryl Hydrocarbon Receptor - Derek V. Henley (Advisor: Mary Ruh) (Willows)
9. The Role of Ras Isoforms and Their Cellular Localization in cc-Thrombin Induced Activation of the ERK and the PI3K Signaling Pathways - Virginia Irintcheva (Advisor: Joe Baldassare) (Willows)
10. Antisense Transport across the Blood-Brain Barrier - Laura Jaeger (Advisor: Bill Banks) (Locust)
11. Modulation of Sympathetic Neurotransmission by NO in the Rat Isolated Perfused Mesenteric Bed - Lacy Kolo (Advisor: Heather Macarthur) (Willows)
12. Stimulation of the Raf/MEK/ERK Cascade is Necessary and Sufficient for Activation and Thr-160 Phosphorylation of a Nuclear-targeted CDK2 - Nathan Lents (Advisor: Joseph Baldassare) (Hawthorne)
13. 5,6-Epoxyeicosatrienoic Acid (5,6-EET) Stimulates Synthesis of Thromboxane A2 in Rabbit Platelets - Jenny Losapio (Advisor: Al Stephenson) (Willows)
14. Molecular Recognition in Tissue Factor-Induced Coagulation - Mathew Ndonwi (Advisor: Paul Bajaj) (Locust)
15. Nitric Oxide Inhibits ATP Release from Erythrocytes - Jeff Olearczyk (Advisor: Randy Sprague) (Willows)
16. SRCAP-Mediated Activation of the Estrogen Response Element - Don Ruhl (Advisor: John Chrivia) (Locust)
17. Thermodynamic Linkage between the SI Site, the Na+ Site, and the Ca2+ Site in the Protease Domain of Human Activated Protein C (APC): Sodium Ion in the APC Crystal Structure is Coordinated to Four Carbonyl Groups from Two Separate Loops - Amy Schmidt (Advisor: Paul Bajaj) (Willows)
18. Danio Rerio: Illuminating the Way to P2X Receptor Function and Physiology - Sarah Stahlhut (Advisor: Mark Voigt) (Hawthorne)
19. Identification of Pathways Involved in Pituitary Lactotroph's Proliferation: Mechanism of Antiproliferative Action of Dopamine - Galina Strizheva (Advisor: Karen Gregerson) (Willows)
20. Central Adrenomedullin's Involvement in the Baroreflex - Meghan Taylor (Advisor: Rick Samson) (Willows)
21. Regulation of CDK4-Cyclin D Activity by Rho A - Charles Travers (Advisor: Joe Baldassare) (Hawthorne)
22. CDK2/Cyclin E Complex Nuclear Translocation - Leroy Wheeler (Advisor: Joe Baldassare) (Locust)

Annual Motivational Retreat

Once a year in the fall, the department would hold an outing in what is referred to as "Wine Country" west of St. Louis. It was held at one of the wineries that grew up along the banks of the Missouri River. These wineries were started by early German settlers. The department would provide the food, but each individual was responsible for their beverage of choice. In most cases the retreat was held around the time that the leaves on the trees were changing to their most colorful. It was usually held on a Friday, and for many years it was a very popular and anticipated event.

"B&B" (The Evening Colloquia)

Another event borrowed from the University of Virginia was "B&B" (which stands for beer and bull). The idea was that one faculty member would be the speaker and another faculty member would be the host and provide snacks and beverages for that evening. It was held at the host's house. Having the event at someone's residence gave it a great deal of informality and atmosphere so that it didn't seem like one was still at work. The charge for the speaker was to talk about new, projected studies that were anticipated or planned—perhaps ideas for a new grant. It was not a seminar. No slides were allowed—only a sketch board. The idea was that the speaker could get input, advice, or criticism that would make the project better. The atmosphere was clearly informal. Faculty, students, and fellows were all invited.

Annual St. Patrick's Day Party

Dr. Westfall and his wife held a St. Patrick's Day Party every year. The highlight of the party was a visit and performance by members of St. Louis Irish Arts. This is an organization started by Dr. P. J. Gannon, professor of psychiatry at the Saint Louis University School of Medicine, and his wife, Helen, who were immigrants from Ireland. St. Louis Irish Arts teaches young children how to play traditional Irish instruments such as the concertina, fiddle, uilleann pipes, Irish whistle, mouth organ, harp, etc., as well as Irish dancing. A group of artists would come to the party and do a performance of musical instruments and dancing. Of course, there were plenty of traditional beverages and a catered supper for all the attendees of the party. All members of the department plus others were invited to this annual celebration.

Dr. Westfall's Annual Dinner Party for Graduate Students

On one of the weekends before Christmas, Dr. Westfall would host a dinner party for all the graduate students and their significant others or guests. Each student or a group of students had to sing a song—a highlight of the party. In addition, for many years, karaoke was part of the entertainment in Dr. Westfall's sports bar. This was an annual event until Dr. Westfall stepped down from the chair in 2013.

Dr. Knuepfer's Octoberfest Party

For many years in late October or early November, Dr. Mark Knuepfer and his wife, Margo, would host an Octoberfest party at their home. There were decorations, music, German sausage, and other German food, including plenty of beer. All members of the department plus neighbors were invited. This was a much-anticipated occasion to honor Dr. Knuepfer's family heritage. For many years Mark's father would travel from Chicago to participate in the festivities.

Spontaneous Celebrations

For many years, once a faculty member was assured of obtaining a research grant from NIH or some other sponsor, he or she would celebrate the occasion by having a party with snacks and beverages. A similar spontaneous party was also carried out once a student passed their preliminary/qualifying exam. Members of the prelim committee and other departmental members were invited. Finally, many birthdays of faculty members or staff were celebrated by members of the department. On some occasions, special birthday greetings were delivered by invited birthday performers (use your imagination). On some occasions when a faculty member moved into a new lab, there were lab-warming parties were held, with gag gifts provided.

The Thomas C. Westfall Graduate Student Symposium

Several faculty, including Mark Knuepfer and Mark Voigt, together with Mary Alice Kauling, Linda

Naes, former graduate Lacy Kolo, and former faculty member Allyn Howlett, organized a graduate student reunion and symposium in honor of Dr. Westfall on the occasion of his stepping down from the chair. The symposium was entitled "The Thomas C. Westfall Graduate Student Symposium." It took place on March 24, 2012, from 8:00 a.m. to 3:45 p.m. in Lecture Hall 3 of the School of Medicine. A copy of the program is enclosed. Over one hundred former graduates, plus current graduate students, former and current faculty, and staff participated. As can be seen in the program, there were fifteen formal presentations, mostly from former students and faculty, plus special presentations from Dr. Westfall's former chairs, Dr. Bill Fleming from West Virginia University and Dr. Joseph Larner from the University of Virginia, in addition to Dr. David Westfall, chair and dean emeritus from the University of Nevada, Reno. The activities began on the evening of March 23 with an informal reception at Dr. Westfall's residence. The formal presentations took place Saturday at the School of Medicine.

The two morning sessions consisted of presentations by five former PhD students who are also faculty members at various institutions: Mark Voigt from Saint Louis University, Katie Gysling from the Catholic University of Chile, Stuart Dryer from the University of Houston, Gonzalo Torres from the University of Pittsburgh, and Sarah Kucenas from the University of Virginia. In addition, there were presentations by two former faculty members and one former postdoctoral fellow: Allyn Howlett, from Wake Forest University; Vincent Chiappinelli, from George Washington University; and Matt Galloway from Wayne State University. The first of two afternoon sessions consisted of former students who chose non-academic career paths: Virginia Wotring at NASA, Melanie Berken at Allegian Medical Affairs, Kevin Gamber

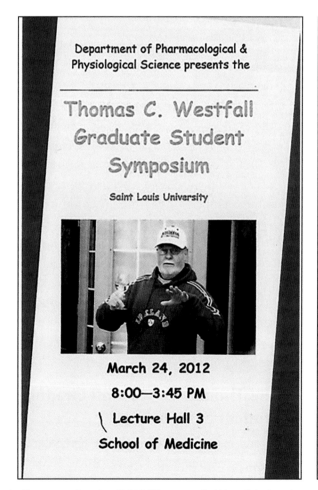

FIG. 29 Program for the 2012 Thomas C. Westfall Graduate Student Symposium

Thomas C. Westfall Graduate Student Symposium Schedule
March 24, 2012

8:00 AM Registration
Continental Breakfast

Former Student Talks: Chair, Alan H. Stephenson, Ph.D.

9:00 Through the Looking Glass: Reductionism, Systems Biology and all Things In Between
Mark M. Voigt, Ph.D. 1981-1985
Professor, Pharmacological and Physiological Science
Saint Louis University School of Medicine

9:20 It Takes Two to Tango with Drugs: CRH-R2 and D1 Receptor Heterodimers in the VTA in Stress-Induced Relapse to Drug Seeking Behaviour
Katia C. Gysling, Ph.D. 1981-1985
Asóciate Profesor y Chair
Departamento Biología Celular y Molecular
Facultad de Ciencias Biológicas
Pontificia Universidad Católica de Chile

9:40 Growth Factors and Developmental Refinement of Neuronal Excitability
Stuart E. Dryer, Ph.D. 1982-1985
John and Rebecca Moores Distinguished Professor
Department of Biology and Biochemistry
University of Houston

10:00 How Does the Brain Manage Dopamine? Lessons from Graduate School
Gonzalo E. Torres, Ph.D. 1994-1999
Associate Professor
Department of Neurobiology
University of Pittsburgh School of Medicine

10:20 Glia: The Other White Meat
Sarah C. Kucenas, Ph.D. 2002-2005
Assistant Professor
Department of Biology
University of Virginia

10:40 AM Coffee Break

Former Faculty and Post-docs: Chair, Mary F. Ruh, Ph.D.

11:00 The Pot-Holder Covers for D2 Receptors in the Basal Ganglia: CB1-D2 Receptor Functional Interactions
Allyn C. Howlett, Ph.D. 1979-2000
Professor, Department of Physiology and Pharmacology
Wake Forest University Health Sciences
Director, Integrative Physiology & Pharmacology Graduate Track
International Cannabinoid Research Society
Board of Directors and Immediate Past President

11:20 Nicotine and the Brain: Where There's Smoke There Are Fired-Up Neurons
Vincent A. Chiappinelli, Ph.D. 1980-1996
Interim Associate Vice Provost for Health Affairs
and Associate Dean of the School of Medicine and Health Sciences
Chair of Pharmacology & Physiology
The Ralph E. Loewy Professor
The George Washington University

11:40 A Novel Approach to Neuropharmacology: Magic Angle Spinning ^{1}H-Magnetic Resonance Spectroscopy
Matthew Galloway, Ph.D. 1980-1982
Departments of Psychiatry & Behavioral Neurosciences, Anesthesiology
Wayne State University

12:00 PM Lunch
PPS Conference Room M360

Afternoon Session

Physiology and Pharmacology Careers: Chair, Roberta J. Secrest, Ph.D.

1:30 Pharmacology during Spaceflight Missions
Virginia E. (Ginger) Wotring, Ph.D. 1991-1997
Space Research Associate
Division of Space Life Sciences
Johnson Space Center Universities

1:45 My Journey to the Dark Side
Melanie S. Berken, Ph.D. 1999-2004
Senior Medical Scientific Manager
Allergan Medical Affairs

2:00 From ARC to Park: Hypothalamic Control of Energy Balance and the Generation of Novel Animal Models
Kevin Gamber, Ph.D. 2000-2005
Global Product Manager Neuroscience Animal Models
Sigma Advanced Genetic Engineering (SAGE) Labs

2:15 Now What? What Happens After You Make the Medical Breakthrough of the Century
Lacy L. Kolo, Ph.D., Esq. 1999-2004
Attorney, Associate Intellectual Property
Patton Boggs LLP

2:30 Pharmacology and Physiology—The catalysts for successful decision making in the life sciences industry
Jennifer (Losapio) Iverson, Ph.D. 2001-2005
Consultant, Commercial Development

2:45 Summary Mark M. Knuepfer, Ph.D.

Afternoon Session

Historical Perspectives: Chair, Mark M. Voigt, Ph.D.

3:00 The Early Years
David P. Westfall, Ph.D.
Former Chair of Pharmacology and Provost
University of Nevada at Reno

3:15 Graduate Training in Pharmacology: the West Virginia University Pharmacology Experience
William W. Fleming, Ph.D.
Former Chair, Department of Pharmacology
West Virginia University

3:30 The Great Years at the University of Virginia
Joseph Larner, Ph.D.
Former Chair, Department of Pharmacology
University of Virginia

Tours of campus/Free time

Reception at SqWires http://sqwires.com/ (314) 865-3522
1415 South 18th St.
Lafayette Square
St. Louis, MO 63104

6:00 PM Cocktails

7:00 PM Dinner

8:00 PM Roast

9:00 PM Entertainment

FIG. 29 Program for the 2012 Thomas C. Westfall Graduate Student Symposium

at Sigma Advanced Genetic Engineering, Lacy Kolo, an associate in intellectual property at Patton Boggs, L.L.P., and Jennifer (Iverson) Losapio in commercial development. The final afternoon session consisted of presentations by David Westfall, chair and dean emeritus at the University of Nevada, Reno; Bill Fleming, Dr. Westfall's former chair at West Virginia University; and Joe Larner, Dr. Westfall's former chair at the University of Virginia. Saturday night there was a celebratory dinner at SqWires restaurant in the Lafayette Square district of St. Louis. Highlights included a roast of Dr. Westfall and numerous speeches and comments from those attending. Figure 30 is a picture of most of the participants. It was a special occasion and one that was sincerely appreciated by Dr. Westfall. One of the highlights of the affair was the announcement of the Thomas C. Westfall Graduate Student Fellowship, an endowment established by contributions from former and current students, faculty, staff, and friends.

Faculty during the Westfall Era (1979–2013)

During the time Dr. Westfall was a department chair there were thirty-seven tenure-track faculty members, (with an average of 17–20 at any one time) seventeen research-track faculty, twenty-three faculty with secondary appointments, and thirteen scientists with adjunct appointments. When Dr. Westfall was appointed chair in 1979, there were three full-time faculty in the department: Dr. Coret, Dr. Kim, and Dr. Gold. From 1979 to 1990, seven new faculty were recruited. These were Dr. Howlett, Dr. Wang, Dr. Chapnick, Dr. Chiappinelli, Dr. Beinfeld, Dr. Knuepfer, and Dr. Andrade. Faculty with secondary

FIG. 30 Attendees: 2012 Thomas C. Westfall Graduate Student Symposium

appointments included Dr. Lonigro, Dr. Poklis, and Dr. Malone. Research-track faculty included Dr. White, Dr. Stephenson, Dr. Strong, Dr. Nemani, and Dr. Meldrum. Following the formation of the Department of Pharmacological and Physiological Science, the following faculty were recruited to the tenure track: Dr. Partridge, Dr. Egan, Dr. Neckameyer, Dr. Chrivia, Dr. MacDonald, Dr. Voigt, Dr. Baldassare, Dr. Samson, Dr. Gautam, Dr. Sprague, Dr. Lonigro, Dr. Gregerson, Dr. Harkins, Dr. Lemon, Dr. Enkvetchakul, and Dr. Salvemini. Full-time faculty previously in the Department of Physiology and who became members of the Department of Pharmacological and Physiological Science were Dr. Senay, Dr. Forrester, Dr. van Beaumont, Dr. Mary Ruh, Dr. Thomas Ruh, Dr. Lechner, Dr. Liu, and Dr. Ellsworth. Upon the reorganization of the Department of Anatomy and Neurobiology in 2004, Dr. Zahm, Dr. Ariel, and Dr. Panneton became members of the Department of Pharmacological and Physiological Science. From 1990 to 2013, research-track faculty included Dr. Quinn, Dr. Han, Dr. Houston, Dr. Cox, Dr. Dong, Dr. Macarthur, Dr. Nemani, Dr. Strong, and Dr. Yoon, and faculty with secondary appointments included Dr. Hertelendy, Dr. Baudendistel, Dr. Moskowitz, Dr. Yoon, Dr. Devaskar, Dr. Li, Dr. Bajaj, Dr. Banks, Dr. Fleisler, Dr. Matuschak, Dr. Willmore, Dr. Gratton, Dr. Kusner, Dr. Zavorsky, Dr. Edwards, Dr. Nowak, Dr. Salvemini, Dr. Martin, and Dr. Malone. During Dr. Westfall's tenure, the following scientists had adjunct appointments: Dr. O'Donohue, Dr. Lathrom, Dr. Wood, Dr. Koepke, Dr. Farah, Dr. Salvemini, Dr. Patton, Dr. Nichols, Dr. Galasko, Dr. Kelce, Dr. Herrmann, Dr. Bond, and Dr. Griggs.

Tenured Faculty, Department of Pharmacology (1979–90)

Original Faculty at the Time of Dr. Westfall's Appointment
(Year of original appointment in parentheses)
- Dr. Irvine Coret (1950)
- Dr. Alvin Gold (1968)
- Dr. Yee Kim (1966)

Faculty Recruited by Dr. Westfall (1979–87)
- Dr. Allyn Howlett (1979)
- Dr. Rex Wang (1979)
- Dr. Barry Chapnick (1979)
- Dr. Vincent Chiappinelli (1980)
- Dr. Margery Beinfeld (1981)
- Dr. Mark Knuepfer (1986)
- Dr. Rodrigo Andrade (1987)

Tenured Faculty, Department of Pharmacological and Physiological Science (1990–2013)

All the above except Dr. Wang.

Faculty Incorporated from the Department of Physiology
- Dr. Charles Senay (1957)
- Dr. Willem van Beaumont (1968)
- Dr. Mary Ruh (1972)
- Dr. Thomas Ruh (1975)
- Dr. Thomas Forrester (1975)
- Dr. Andrew Lechner (1981)
- Dr. Maw-Shung Liu (1982)
- Dr. Mary Ellsworth (1988)

Tenure-Track Faculty Recruited by Dr. Westfall
- Dr. Nicola Partridge (1991)
- Dr. Terrance Egan (1992)
- Dr. Wendi Neckameyer (1992)
- Dr. Paul MacDonald (1993)
- Dr. John Chrivia (1993)
- Dr. Mark Voigt (1994)
- Dr. Joseph Baldassare (1996)
- Dr. Willis "Rick" Samson (1999)
- Dr. Medha Gautam (1997)
- Dr. Randy Sprague (2000)
- Dr. Andrew Lonigro (2000)
- Dr. Heather Macarthur (1997)
- Dr. Al Stephenson (2007)
- Dr. Karen Gregerson (2001)
- Dr. Amy Harkins (2002)
- Dr. Christian Lemon (2008)
- Dr. Decha Enkvetchakul (2008)
- Dr. Daniela Salvemini (2009)

Faculty Obtained from Anatomy and Neurobiology
- Dr. Daniel "Scott" Zahm (1985)
- Dr. Michael Ariel (1993)
- Dr. Michael Panneton (1981)

Secondary, Research, and Adjunct Appointments during Dr. Westfall's Tenure as Chair

Research Track Faculty

- Dr. Francis White
- Dr. Al Stephenson
- Dr. Randy Strong
- Dr. Ramakrishna Nemani
- Dr. Cheryl Quinn
- Dr. Song-ping Han
- Dr. Devin Houston
- Dr. Jane Cox
- Dr. Lin-Wang Dong
- Dr. Heather Macarthur
- Dr. Gina Yosten
- Dr. Timothy Doyle
- Dr. Alice Gardner
- Dr. Michael Meldrum
- Dr. John Martin
- Dr. Helen McIntosh

Secondary Faculty Appointments

- Dr. Randy Strong
- Dr. Andrew Lonigro
- Dr. Alphonse Poklis
- Dr. David Malone
- Dr. Frank Hertelendy
- Dr. Lawrence Baudendistel
- Dr. David Moskowitz
- Dr. Kong-Woo Peter Yoon
- Dr. Sherin Devaskar
- Dr. Albert P. Li
- Dr. Rama Nemani
- Dr. Felicia V. Nowak
- Dr. Paul Bajaj
- Dr. William Banks
- Dr. Steven Fleisler
- Dr. George Matuschak
- Dr. L. James Willmore
- Dr. Michael Anne Gratton
- Dr. Linda Kusner
- Dr. Gerald Zavorsky
- Dr. John Edwards
- Dr. Daniela Salvemini
- Dr. Scott Martin

Adjunct Appointments

- Dr. Thomas O'Donohue
- Dr. Thomas H. Lanthorn
- Dr. Paul L. Wood
- Dr. John Koepke
- Dr. John Farah
- Dr. Kevin Patton
- Dr. D. Allen Nichols
- Dr. Gail T. F. Galasko
- Dr. William R. Kelce
- Dr. Daniela Salvemini
- Dr. Steven Herrmann
- Dr. Brian R. Bond
- Dr. David W. Griggs

Faculty during the Westfall Era, Part II

(Years of service at Saint Louis University in parentheses.)

Nicola C. Partridge, PhD (1991–2000)

Dr. Partridge is a native of Perth, Australia. She earned a BSc (Hons I) in biochemistry in 1973 and a PhD in biochemistry in 1981, both from the University of Western Australia. She was a postdoctoral fellow in the laboratory of Dr. T. John Martin at the University of Melbourne from 1981 to 1984. From 1984 to 1985, she had a position in the Division of Cell Biology and Biochemistry at the Washington University School of Dental Medicine, followed by a faculty position in the Department of Pediatrics at the Saint Louis University School of Medicine from 1985 to 1991. She accepted a position of associate professor in the newly formed Department of Pharmacological and Physiological Science in 1991 and was promoted to professor in 1996. Dr. Partridge also held a secondary position in the Department of Orthopedic Surgery and was an active member of the Cell and Molecular Biology graduate training program. Dr. Partridge's research interests were in the regulation of gene expression of osteoblasts, signaling transduction pathways mediating hormone action, and the role of metalloproteinases in bone. She was well funded both by the NIH and the National Aeronautics and Space Administration (NASA) and had several projects carried to orbit on space shuttles. Dr. Partridge was codirector of the CORE Graduate Program in Basic Biomedical Sciences until a permanent director was recruited in 1999. In 2000 Dr. Partridge accepted the position of chairperson of the Department of Physiology at the Robert Wood Johnson School of Medicine of the State University of New Jersey. From 2007 to 2008, she served as president of the Society of Physiology Chairs. In 2009 she accepted the chair of Biomedical Sciences at the New York University School of Dental Medicine, where she is continuing her outstanding career and leadership role.

Terrance M. Egan, PhD (1992–present)

Dr. Egan is a native of upstate New York. He earned a BSc in pharmacy from the Creighton University

School of Pharmacy in Omaha in 1978. After working in the Department of Pharmacology at the Loyola University School of Medicine, he transferred with his PhD advisor, Dr. R. Alan North, and earned his PhD in neural and endocrine control at the Massachusetts Institute of Technology (MIT) in 1984. Dr. Egan did postdoctoral studies with Professor Denis Noble at the University Laboratory of Physiology, University of Oxford, from 1985 to 1987; with Professor Hans Dieter Lux at Abteilung Neurophysiologie, Max-Planck-Institut für Biochemie, in Germany from 1987 to 1988; with Professor Levitan in the Graduate Department of Biochemistry, Brandeis University, from 1988 to 1991, and with Professors W. Gil Wier and C. William Balke in the Department of Cardiology and Physiology at the University of Maryland Medical School. Dr. Egan accepted a position as assistant professor in the Department of Pharmacological and Physiological Science in 1992. He was promoted to associate professor in 1999 and professor in 2004. Dr. Egan was a visiting research fellow in the MRC Laboratory of Molecular Biology with Dr. Baljit Singh Khakh at Cambridge University from 2002 to 2003. Dr. Egan's research has been in the control of ion channel activity using electrophysiological and molecular biological approaches. He is an internationally recognized expert on purinergic receptors, especially P2X subtypes. Dr. Egan has received numerous grants from the NIH and the American Heart Association and has been a member of several permanent NIH study sections. He has served as director of graduate studies of the pharmacology/physiology training program and co-program director of a T32 training grant in Pharmacological Sciences. He is currently actively continuing his outstanding research and scholarly career in the Department of Pharmacology and Physiology. In his spare time, Dr. Egan runs numerous marathons, including in Boston and New York.

Wendi Neckameyer, PhD (1992–2016)

Dr. Neckameyer is a native of New York City. She earned a BS in biochemistry from Cornell University in 1980 and a PhD in viral oncology from Rockefeller University in 1985 under the tutelage of Dr. Lu-Hai Wang. She did postdoctoral studies in the laboratory of Dr. William G. Quinn from 1985 to 1988 in the Department of Brain and Cognitive Sciences and Department of Biology at the Massachusetts Institute of Technology. Dr. Neckameyer was a senior research associate in the Department of Biology at Brandeis University from 1988 to 1992, working with Dr. Kalpana White. Dr. Neckameyer joined the Department of Pharmacological and Physiological Science at the Saint Louis University School of Medicine as an assistant professor in 1992. She was promoted to associate professor in 1999 and professor in 2008. In 1999 Dr. Neckameyer was a visiting professor at the Université de Paris Sud in Orsay. Dr. Neckameyer's research was in *Drosophila* molecular neurobiology and molecular genetics and the role of tyrosine hydroxylase and tryptophan hydrolysis in behavior. She received funding from both the NIMH and the NSF and served on review panels for both organizations. Dr. Neckameyer left the department in 2016.

John Chrivia, PhD (1993–present)

Dr. Chrivia is a native of California. He earned a BS in biochemistry in 1980 from the University of California, Davis. This was followed by a PhD in pharmacology from the University of Washington in 1987 under the mentorship of Dr. G. Stanley McKnight. Dr. Chrivia did postdoctoral studies at the Baylor University School of Medicine from 1988 to 1990 and the Vollum Institute for Advanced Biomedical Research at the Oregon Health & Sciences University in the laboratory of R. H. Goodman from 1990 to 1993. Dr. Chrivia accepted a position as an assistant professor in the Department of Pharmacological and Physiological Science at Saint Louis University in 1993. He was promoted to associate professor in 2000 and professor in 2005. Dr. Chrivia's research is in molecular endocrinology and cancer biology, concentration transcription factors, coactivation, and nuclear proteins. He has received funding from the NIH, the American Heart Association, and the US Department of Defense. More recently, Dr. Chrivia is concentrating on teaching medical, graduate, and undergraduate students and has served as chair of the Animal Care Committee in the Department of Pharmacology and Physiology, first under the chairmanship of Dr. Thomas Burris followed by those of Dr. Mark Voigt and now Dr. Daniela Salvemini.

Paul MacDonald, PhD (1993–99)

Dr. MacDonald is a native of Erie, Pennsylvania. He earned a BS in chemistry at Edinboro University in Pennsylvania in 1984 and a PhD in biochemistry from Vanderbilt University in 1988. He was a postdoctoral research associate in the Department of Biochemistry at the University of Arizona from 1989 to 1992 and a research assistant professor of medicine in the Renal Division of the Washington University School of Medicine from 1992 to 1993. Dr. MacDonald joined the Department of Pharmacological and Physiological Science at Saint Louis University as assistant professor in 1994. He was promoted to associate professor in 1998. Dr. MacDonald's research involved molecular endocrinology, especially the regulation of vitamin D and estrogen receptors. His research was well funded by the NIH and the Osteoporosis Society. He was recruited to the Department of Pharmacology at the Case Western Reserve University School of Medicine as director of the Predoctoral Training Program in Molecular Therapeutics. He is continuing his outstanding research as professor and program director of a T32 in molecular therapeutics.

Mark Massie Voigt, PhD (1994–2019)

Dr. Voigt is a native of St. Louis County. He earned a BA in chemistry from Central Methodist College in Fayette, Missouri, in 1979. This was followed by an MS in biochemistry at the University of Missouri, Columbia in 1981 and a PhD in pharmacology under the tutelage of Dr. Thomas C. Westfall at Saint Louis University in 1985. He did postdoctoral studies at the Howard Hughes Medical Institute of the Department of Neurology at Massachusetts General Hospital and Harvard University Medical School under the supervision of Dr. G. Uhl; the Laboratory of Molecular Biology and Laboratory Division of Neurobiology at NINDS and the National Institute of Child Health and Human Development (NICHD) under the supervision of Dr. Richard Henneberry and Dr. Philip Nelson; and finally as an Alexander von Humboldt Fellow in the Laboratory of Molecular Neuroendocrinology at the University of Heidelberg in Germany with Dr. Peter Seeburg. From 1991 to 1994, Dr. Voigt was a college lecturer in the Department of Pharmacology and Biotechnology Institute at University College Dublin in Ireland. Dr. Voigt returned to the United States as assistant professor in the Department of Pharmacological and Physiological Science at Saint Louis University in 1994. He was promoted to associate professor in 1999 and professor in 2005. Dr. Voigt has carried out research on the molecular and functional analysis of 5-hydroxytryptamine and purinergic receptors and neuronal development of the latter using zebrafish as a model. His research has been well supported by the NIH, the National Association for Research on Schizophrenia and Depression (NARSAD), and NATO. In addition to his outstanding research and teaching, Dr. Voigt has served numerous leadership roles at Saint Louis University, including vice chair under both Dr. Westfall and Dr. Burris, interim chair of the Department of Pharmacological and Physiological Science from January to July 2013 and again in 2018, associate dean of research at Doisy College of Health Sciences from 2013 to 2015, and director of the Center for Neuroscience. He also served on important committees both at Saint Louis University and nationally and internationally including being Co-Program Director of a T32 training grant in pharmacological sciences. Dr. Voigt continued his exceptional research, teaching, and service until his retirement in 2019.

Joseph Baldassare, PhD (1996–present)

Dr. Baldassare is a native of Boston. He attended both Williams College and Boston College, where he earned a BS in physics in 1964. He did graduate work at the Massachusetts Institute of Technology (MIT) and the University of Pittsburgh, where he earned a PhD in biophysics in 1970. He was a research associate in the Department of Biological Chemistry at the Washington University School of Medicine from 1974 to 1980 and an instructor in the Department of Biological Chemistry at Washington University from 1980 to 1983. He was a research scientist at American Red Cross Blood Services in St. Louis from 1987 to 1989. He accepted a position as associate professor of medicine in the Department of Internal Medicine at the Saint Louis University School of Medicine from 1990 to 1995 and was promoted to professor in the Department of Internal Medicine at SLU in 1995. In 1996 Dr. Baldassare accepted the position of professor in the Department of Pharmacological and Physiological Science at the Saint Louis University School of Medicine. Dr. Baldassare's research is in the area of G protein-coupled receptors and signal transduction, and he is an internationally recognized authority on cell cycling in human platelets. He has had long-lasting funding from the NIH and has mentored numerous PhD students. He has been a longtime director of graduate studies in the Department of Pharmacological and Physiological Science. He is an outstanding teacher of medical, graduate, and undergraduate students and is much sought after as an educator. He is currently continuing his activity as an excellent teacher in phased retirement in the Department of Pharmacology and Physiology.

Heather Macarthur, PhD (1997–present)

Dr. Macarthur was raised in the village of Lochcarron near the Isle of Skye in the highlands of northwest Scotland. She earned a BSc in pharmacology at the University of Edinburgh in 1990 and a PhD in pharmacology in 1994 at the William Harvey Research Institute at Saint Bartholomew's Medical College in London. Heather's mentor was Professor Sir John Vane, FRS, who was awarded the Nobel Prize for Physiology or Medicine in 1982 for his discoveries concerning prostaglandins and related biologically active substances. From 1994 to 1997, Dr. Macarthur was a postdoctoral research fellow in the laboratory of Dr. Thomas Westfall in the Department of Pharmacological and Physiological Science at the Saint Louis University School of Medicine, working on a postdoctoral research fellowship from the Missouri division of the American Heart Association. She worked as an assistant research professor in the Department of Pharmacological and Physiological Science from 1997 to 2004 under the chairmanship of Dr. Westfall, as well as a visiting research scientist at C. G. Searle and Co. (now Pfizer, Inc.) from 1997 to 1999. She was promoted to associate research professor in 2004 and appointed to the tenure track in 2008 as associate professor in the Department of Pharmacological and Physiological Science. Dr. Macarthur's research interests include vascular control mechanisms, endothelial mediators, regulation of a sympathetic neurotransmission, and the pathophysiology of Parkinson's disease. She made the important discovery that during septic shock nitric oxide chemically inactivates catecholamines, contributing to hypotension and shock. Her research has been supported by the NIH, the American Heart Association, and private industry. Dr. Macarthur is an outstanding teacher and has mentored numerous PhD students. She has also served on many school and university committees and is currently course master of the principles of pharmacology module for medical students. Dr. Macarthur continued her exciting career in the Department of Pharmacology and Physiology under the chairmanship of Dr. Thomas Burris and currently works under the interim chairmanship of Dr. Daniela Salvemini.

Alan H. Stephenson, PhD (1987–2016)

Dr. Stephenson is a native of Wisconsin. He earned a BS in biology and chemistry from Carroll College in Waukesha, Wisconsin, in 1971. This was followed by an MS in zoology (comparative physiology) at the University of Wisconsin-Milwaukee in 1974 and a PhD in pharmacology at Saint Louis University in 1985 under the mentorship of Dr. Andrew Lonigro. Dr. Stephenson was a postdoctoral research associate in the Department of Medicine at Saint Louis University from 1985 to 1986 and assistant research professor in the Department of Pharmacology from 1986 to 1989 under the chairmanship of Dr. Thomas C. Westfall. He was promoted to associate research professor in 1989 and research professor in 1998 in the Department of Pharmacological and Physiological Science. He was appointed to professor, tenure track, in 2007. Dr. Stephenson carried out outstanding research on vascular control mechanisms and pulmonary and cardiovascular pharmacology and was a world authority on the physiology and pharmacology of eicosanoids. Dr. Stephenson's research was continuously supported by the NHLBI and the American Heart Association. He mentored numerous PhD students, medical students, and postdoctoral fellows. He was also an outstanding teacher and served on numerous department, university, and national committees. He was vice chair and chair of the Institutional Animal Care and Use Committees (IACUC) and served on American Heart Association and NIH study sections. Dr. Stephenson retired in 2016 and is currently professor emeritus in the Department of Pharmacology and Physiology at Saint Louis University.

Randy Stephen Sprague, MD (2000–16)

Dr. Sprague is a native of Belleville, Illinois, and grew up in the St. Louis area. He earned a BA in biology from the University of Kansas in 1971, followed by an MD at the Saint Louis University School of Medicine in 1976. He completed his internship and residency at Saint Louis University Hospital and was chief resident in 1978–79. He was a fellow in pulmonary medicine at the Saint Louis University School of Medicine and attending physician at the St. Louis VA Hospital and Saint Louis University Hospital. He was appointed assistant professor in the Department of Internal Medicine in 1981 and was promoted to associate professor in 1989 and professor in 1997. Dr. Sprague was a visiting scientist at the William Harvey Research Institute at St. Bartholomew's Medical College in London from 1991 to 1992, working with Nobel laureate Sir John Vane. Dr. Sprague was appointed professor in the Department of Pharmacological and Physiological Science at the Saint Louis University School of Medicine in 1999. Dr. Sprague's research is in the area of respiratory and cardiovascular physiology, and he is a world authority on ATP release from RBCs. He has been well funded by the NIH and the American Diabetes Foundation. Dr. Sprague is an outstanding teacher and mentor and has won numerous teaching awards from medical students at Saint Louis University, including the Golden Apple Award in 1997 and 1998, Teacher of the Year in 1991 and 1993, and the Osler Award for Teaching Excellence in 1987. He has mentored numerous PhD and MD/PhD students. He also has performed important leadership roles at Saint Louis University, including being chairman of the Animal Care Committee and president of the Faculty Senate. Dr. Sprague retired in 2016 but continues to be active in teaching and service as professor emeritus in the Department of Pharmacology and Physiology and the Department of Internal Medicine.

Willis Kendrick "Rick" Samson, PhD, DSc (Hons) (1999–present)

Dr. Samson is a native of upstate New York. He earned a BA in chemistry from Duke University in 1968 and an Interpreter's Diploma in German at the Defense Language Institute in Monterey, California, in 1970. From 1969 to 1972, Dr. Samson served in the United States Army, then earned his PhD in physiology from the University of Texas Health Science Center (UTHSC) 1979 under the mentorship of S. M. McCann. Dr. Samson was awarded a DSc (Hons) from Westminister College in Fulton, Missouri, in 2010. He did postdoctoral training at UTHSC from 1979 to 1981 and was assistant professor of physiology there from 1981 to 1988. He was an associate professor in the Department of Anatomy and Neurobiology at the University of Missouri School of Medicine from 1988 to 1992. In 1992 Dr. Samson was appointed professor and chairman of the Department of Physiology at the University of North Dakota School of Medicine, a position he held until 1999. From 1998 to 1999, he also held the position of acting chairman of the Department of Pharmacology and Toxicology at the University of North Dakota. Dr. Samson accepted the positions of director of the Core Graduate Program in Biomedical Sciences and professor in the Department of Pharmacological and Physiological Science at Saint Louis University School of Medicine in 1999. In 2009, Dr. Samson was also named director of graduate programs in the biomedical sciences at the same institution. In addition to his leadership role with the Core Graduate Program, Dr. Samson is an active investigator in neuroendocrinology and is an internationally recognized expert in the hypothalamic control of anterior pituitary function, the physiology of vasoactive peptides, and the CNS and hormonal control of cardiovascular function and control of salt and water appetite. Dr. Samson has had several leadership roles for the American Physiology Society, the American Heart Association, and the NIH, including serving as editor in chief of the *American Journal of Physiology: Regulatory, Integrative, and Comparative Physiology*. His research has been continuously funded by the American Heart Association and the NIH. Dr. Samson is continuing his outstanding career as professor in the Department of Pharmacology and Physiology, first under the chairmanship of Dr. Thomas Burris and currently under Dr. Daniela Salvemini. He continues to work as the director of the Core Graduate Program in Biomedical Sciences.

Medha Gautam, PhD (1997–2006)

Dr. Gautam is a native of India. She earned a BS in chemistry from St. Xavier's College in Mumbai, India, in 1979. This was followed by an MS in biochemistry from the Topiwala National Medical College in Mumbai in 1981 and a PhD in molecular biology at the Tata Institute of Fundamental Research in 1985. She was a research fellow in the Division of Biology at the California Institute of Technology and then a senior research fellow in the Department of Molecular Biology and Pharmacology at the Washington University School of Medicine from 1981 to 1995. She was a research assistant professor in that same department from 1995 to 1997, at which time she joined the Department of Pharmacological and Physiological Science at the Saint Louis University School of Medicine as assistant professor. Dr. Gautam is a molecular and developmental neuroscientist with expertise in proteins and receptors involved in the development of the neuromuscular junction. She accepted a position in basic science at the Southern Illinois University School of Dentistry in 2006, where she has continued her career.

Andrew Lonigro, MD (2000–07)

Dr. Lonigro was a native of St. Louis. He received a BS from Saint Louis University in 1958. He followed this with studies from 1959 to 1962 at the University of Heidelberg, where he received certificates for proficiency in German and completion of the first examination in medicine. He earned his MD from Saint Louis University School of Medicine in 1966. Dr. Lonigro was a resident in clinical medicine at Saint Louis University Hospital from 1969 to 1971 and a cardiology research fellow those same years. He was associate attending staff in the Department of Medicine and chief of hypertension at the VA Hospital and Medical College of Wisconsin in Milwaukee from 1971 to 1976. He was a staff physician at Cardinal Glennon Children's Hospital in St. Louis, chief of clinical pharmacology at the St. Louis VA Hospital, and director of the Division of Clinical Pharmacology at Saint Louis University Hospital. From 1971 to 1976, Dr. Lonigro was also an assistant professor in the Departments of Medicine and Pharmacology at the Medical College of Wisconsin. He was associate professor from 1976 to 1984 and professor from 1984 to 2007 in the Department of Medicine at the Saint Louis University School of Medicine, with a secondary appointment in the Department of Pharmacology. He was appointed professor in the Department of Pharmacological and Physiological Science, with a secondary appointment in the Department of Medicine at the same institution in 2000 and served until his death in 2007. During this period, he also served as director of the Division of Clinical Pharmacology. Dr. Lonigro was an internationally recognized authority on the role of prostaglandins in the cardiovascular system, especially in the pulmonary system and in pericytes. His research was well funded by the NIH, the American Heart Association, and the Veterans Administration. He also served on numerous review committees in these organizations, including as chair of the Cardiovascular and Renal Study Section at NIH. Dr. Lonigro also served on numerous committees for the Saint Louis University School of Medicine, Saint Louis University Hospital, the local American Heart Association, and the Veterans Administration. He also mentored numerous MD, PhD, and MD/PhD fellows and students and was a popular professor with medical students. Dr. Lonigro was active in research, teaching, and service until his untimely death in 2007.

Karen Ann Gregerson, PhD (2001–04)

Dr. Gregerson earned a BA in biology from Illinois Wesleyan University in 1976 and a PhD in physiology from the University of Nebraska School of Medicine in 1981, working with Dr. Gary Campbell. She was a postdoctoral fellow in the Department of Physiology at the University of Maryland School of Medicine, working with Dr. Michael Selmanoff, from 1982 to 1985 and at the Department of Physiology at the University of North Carolina at Chapel Hill with Dr. Garry Oxford in 1986. She was appointed assistant professor of pediatrics and physiology in 1987 and associate professor in those same departments in 1995, serving there until 1998. She became associate professor of obstetrics/gynecology and reproductive sciences and associate professor of physiology from 1996 to 2001 at the University of Maryland School of Medicine. She accepted a position as associate professor in the Department of Pharmacological and Physiological Science at Saint Louis University School of Medicine in 2001. Dr. Gregerson's research was in the regulation of prolactin secretion and endocrine physiology. She accepted a position in the Department of Physiology at the University of Cincinnati in 2004, where she continued her work in endocrine physiology until her untimely death in 2014.

Amy B. Harkins, PhD (2002–present)

Dr. Harkins is a native of Texas. She earned a BA in biology at the University of Texas, Austin in 1986, followed by an MS in neuroscience at the University of Texas, San Antonio in 1988 and a PhD in neuroscience at the University of Pennsylvania under the tutelage of Dr. Stephen Baylor in 1993. Dr. Harkins was a postdoctoral fellow in the laboratory of Dr. Aaron Fox in the Department of Pharmacological and Physiological Science at the University of Chicago from 1993 to 1997. She remained at Chicago as a research associate from 1997 to 2000 and an assistant professor from 2000 to 2002. She accepted a position as assistant professor in the Department of Pharmacological and Physiological Science at the Saint Louis University School of Medicine in 2002 and was promoted to associate professor in 2009. Dr. Harkins was a visiting professor on sabbatical leave in the Department of Biomedical Engineering at Washington University in St. Louis in 2012. Dr. Harkins's research interests are in the regulation of synaptic transmission, neuronal signaling, and neuronal regeneration in 3D scaffolds, biomaterials, and bioanalytical chemistry. Dr. Harkins has been active in the Biophysics Society and holds a secondary appointment in the Department of Biomedical Engineering at Saint Louis University. She has received extramural funding from numerous sources, including the NIH, the NSF, and the Whitehall Foundation, among others. Dr. Harkins has been particularly active in mentoring undergraduate students, high school students, and graduate students and has received a Five-Year Service Award for the NSF research mentor program Supporting Talent for Academic Recruitment in STEM (STARS). Dr. Harkins is currently serving as chair of the Department of Health Sciences and Informatics at the Doisy College of Health Sciences.

Decha Enkvetchakul, MD (2008–16)

Dr. Enkvetchakul is a native of Missouri. He earned his BA in biology at Washington University in St. Louis in 1989 and an MD from the University of Missouri School of Medicine in 1993. He did a residency in internal medicine at Barnes Hospital from 1993 to 1996 and a clinical fellowship in nephrology at Barnes-Jewish Hospital from 1997 to 2001. He simultaneously did a research fellowship at the Washington University School of Medicine, working in the laboratory of Dr. Colin Nichols, from 1997 to 2001. Dr. Enkvetchakul was an instructor in internal medicine at the Washington University School of Medicine from 2001 to 2006, which he followed with a research assistant professorship in the Department of Cell Biology and Physiology at the Washington University School of Medicine from 2006 to 2008. Dr. Enkvetchakul was appointed assistant professor in the Department of Pharmacological and Physiological Science at the Saint Louis University School of Medicine in 2008, where he remained until his departure in 2016. Dr. Enkvetchakul was a popular teacher with medical students. His research was in the area of renal physiology and the structure and function of KATP channels especially KirBac1.1 channel activity. Dr. Enkvetchakul left Saint Louis University in 2016 to pursue other endeavors and is currently in private practice in the St. Louis area.

Christian H. Lemon, PhD (2008–13)

Dr. Lemon is a native of Oklahoma. He earned a BS in psychology from the University of Oklahoma in 1994. This was followed by an MA in psychology at the State University of New York at Binghamton in 1998 and a PhD in psychobiology at that same institution in 2001. Dr. Lemon was a postdoctoral fellow in the Department of Anatomy and Neurobiology at the University of Maryland School of Medicine from 2001 to 2002, working under the tutelage of Dr. David Smith. He accompanied Dr. Smith as a postdoctoral fellow to the Department of Anatomy and Neurobiology at the University of Tennessee Health Science Center in 2002 when Dr. Smith assumed the chairmanship of that department. From 2004 to

2007, Dr. Lemon served as a research assistant professor there until he was appointed assistant professor in the Department of Pharmacological and Physiological Science at the Saint Louis University School of Medicine in 2008. Dr. Lemon is a neuroscientist; his research is in the area of gustatory neural coding and sensory perception. Dr. Lemon accepted a position in the Department of Neuroscience at the University of Oklahoma in 2013.

Daniela Salvemini, PhD (2009–present)

Dr. Salvemini was born in Africa to Italian parents and educated in the United Kingdom. She earned a BS in pharmacology in 1987 at King's College London and a PhD in pharmacology at the William Harvey Research Institute at St. Bartholomew's Medical College under the tutelage of Nobel laureate Sir John Vane in 1990. Following a postdoctoral fellowship with Dr. Vane from 1990 to 1992, Dr. Salvemini continued her postdoctoral training in the Department of Molecular Pharmacology at Monsanto Research in St. Louis, working in the laboratory of Dr. Philip Needleman from 1992 to 1994. In 1994 Dr. Salvemini moved to G.D. Searle Discovery Research, where she was a senior research investigator from 1994 to 1995, a research scientist I from 1995 to 1997, and a research scientist II and project leader from 1997 to 1999. From 1999 to 2005, Dr. Salvemini worked at Metaphore Pharmaceuticals, Inc., where she was vice president of biological and pharmacological research from 2001 to 2003 and senior vice president of research from 2003 to 2005. From 1996 to 2006, Dr. Salvemini was an adjunct faculty member in the Department of Pharmacological and Physiological Science at the Saint Louis University School of Medicine. She accepted a position as research professor in internal medicine there in 2005 and also held a secondary position as research professor in the Department of Pharmacological and Physiological Science. Dr. Salvemini was appointed associate professor of the Department of Pharmacological and Physiological Science in 2009 and was promoted to professor in 2011. She has been an adjunct professor in the Molecular Pharmacology Faculty of Pharmacy in Catanzaro, Italy, since 2000 and is a visiting professor in the Department of Preclinical and Clinical Pharmacology at the University of Florence in Italy.

Dr. Salvemini is an internationally recognized scholar and investigator, and her research efforts have provided the pharmacological rationale for therapeutic development of novel approaches for the management of disease states governed by overt production of oxygen/nitrogen species, sphingolipids, and loss of adenosinergic signaling; autoimmune, inflammatory, cardiovascular, and neurological diseases; cancer; and pain. She has published numerous peer-reviewed papers, been invited to speak at many international meetings, and has received impressive extramural support from the NIH and other funding agencies. Dr. Salvemini has been awarded eleven high-profile awards for excellence in research including the Novartis Prize in Pharmacology; the Premio Internazionale Maria Luisa de' Medici Award, given in Italy to a woman whose distinguished work is of international interest; the St. Louis Academy of Science Outstanding Scientist Award; a fellow of the National Academy of Inventors and the 2020 Pharmacia-ASPETAward for Experimental Therapeutics. Dr. Salvemini has also mentored numerous PhD and MD/PhD students and served on many local, national, and international committees. She is the last faculty member recruited under Dr. Westfall's chairmanship. Dr. Salvemini continued her outstanding career in the Department of Pharmacology and Physiology under the chairmanships of Dr. Thomas Burris and Dr. Mark Voigt and is currently the director of the Henry and Amelia Nasrallah Center for Neuroscience and also the Interim Chair of the Department of Pharmacology and Physiology.

Daniel Scott Zahm, PhD (2004–present)

Dr. Zahm is a native of Lebanon, Pennsylvania. After attending Harrisburg Area Community College, he earned a BS in biology at Bloomsburg University in 1977. This was followed by a PhD in anatomy at the Milton S. Hershey Medical Center at Pennsylvania State University in 1982. Dr. Zahm did a neuroscience postdoctoral fellowship in the laboratory of Lennart Heimer at the University of Virginia from 1982 to 1985. He was appointed assistant professor in the Department of Anatomy and Neurobiology at the Saint Louis University School of Medicine in 1985. He was promoted to associate professor in 1989 and professor in 1995. With the reorganization of the Department of Anatomy and Neurobiology in 2004, Dr. Zahm was appointed professor in the Department of Pharmacological and Physiological Science in Saint Louis. In 1997 he was the Forbairt Visiting Professor at the Royal College of Surgeons and Trinity University in Dublin, Ireland, and the Leverhulme Visiting Professor at the University of St. Andrews in Scotland in 2000. Dr. Zahm is an internationally recognized expert on the anatomy and function of the limbic system, the extended amygdala, and the nucleus accumbens, as well as the neurobiology of neurotensin and substance abuse. His research has been continuously funded by the NIH, and he has published numerous peer-reviewed papers and book chapters. He is an excellent teacher and mentor of both predoctoral and postdoctoral fellows and has served on multiple NIH study sections and committees, both locally and nationally. He has organized four national/international symposia and has been an invited speaker at numerous national/international meetings and universities. Dr. Zahm has continued his outstanding career in the Department of Pharmacology and Physiology under the chairmanships of Dr. Thomas Burris and currently Dr. Daniela Salvemini.

Michael Ariel, PhD (2004–Present)

Dr. Ariel earned a BS in life science and psychology from the Massachusetts Institute of Technology in 1976 and a PhD in neural science from Washington University in St. Louis in 1976. This was followed by postdoctoral training in the Department of Biology at Harvard University from 1980 to 1983 and a position as research associate at the Eye Research Institute at the Retina Foundation from 1993 to 1995. Dr. Ariel was appointed assistant professor in the Department of Behavioral Neuroscience at the University of Pittsburgh from 1985 to 1993. He accepted a position as assistant professor in the Department of Anatomy and Neurobiology at the Saint Louis University School of Medicine in 1993. He was promoted to associate professor in 1996 and professor in 2003. He was a visiting professor on sabbatical leave with Dr. Yosef Yarom in the Neurobiology Department at Hebrew University in Jerusalem in 2000–01 and again in 2009–10. Upon the reorganization of the Department of Anatomy and Neurobiology in 2004, Dr. Ariel was appointed professor in the Department of Pharmacological and Physiological Science. Dr. Ariel is an internationally recognized authority on the visual system and the neuroanatomy and neurophysiology of the ocular system of the turtle. His research has been well funded by the NIH. He is an excellent teacher and mentor and has served on numerous local and national committees. Dr. Ariel continued his outstanding academic career in the Department of Pharmacology and Physiology under the chairmanships of Dr. Thomas Burris, Dr. Mark Voigt and now Dr. Daniela Salvemini.

William Michael Panneton, PhD (2004–2015)

Dr. Panneton was born in Pittsburgh, Pennsylvania. He earned a BS in biology in 1971 from Georgetown University, followed by a PhD in anatomy from Ohio State University in 1978 under the mentorship of Dr. George F. Martin. He did postdoctoral studies at the Washington University School of Medicine in neurobiology, working with Arthur Loewy and Harold Burton from 1978 to 1981. He was appointed assistant professor in the Department of Anatomy and Neurobiology at the Saint Louis University School of Medicine in 1981 and was promoted to associate professor in 1989 and professor in 1995. With the reorganization of the Department of Anatomy and Neurobiology in 2004, Dr. Panneton was appointed professor in the Department of Pharmacological and Physiological Science. Dr. Panneton's research was mainly in understanding the neuroanatomy of central autonomic control areas, their projections, the diving reflex, and the role of DOPAL in Parkinson's disease. His research was funded by the NIH, various foundations, and private industry. He served on numerous university committees, often in leadership roles, and mentored many undergraduate, graduate, and medical students. Dr. Panneton retired in 2015.

Alvin Gold, PhD
See page 42.

Yee Kim, PhD
See page 41.

Allyn Howlett, PhD
See page 78.

Barry Chapnick, PhD
See page 79.

Vincent Chiappinelli, PhD
See page 79.

Margery Beinfeld, PhD
See page 79.

Mark Knuepfer, PhD
See page 80.

Rodrigo Andrade, PhD
See page 81.

Leo C. Senay Jr., PhD
See page 50.

Karol Willem van Beaumont, PhD
See page 59.

Mary F. (Christiano) Ruh, PhD
See page 58.

Thomas S. Ruh, PhD
See page 58.

Thomas Forrester, PhD, MD
See page 59.

Andrew John Lechner Jr., PhD
See page 60.

Maw-Shung Liu, DDS, PhD
See page 59.

Mary L. Ellsworth, PhD
See page 60.

PhD Graduates during the Westfall Era (1979–2013)

During the time Dr. Westfall was chair of the Department of Pharmacology and then the Department of Pharmacological and Physiological Science, there was a flourishing and highly productive graduate program that produced approximately four PhD graduates per year. During the period, a total of 126 PhDs were awarded to students mentored by faculty in the department; sixteen of these occurring in the Department of Pharmacology from 1979 to 1990, and the remainder falling after the formation of the Department of Pharmacological and Physiological Science, from 1991 to 2017.

Financial support for the students came from a variety of sources. These included funds from the School of Medicine; individual faculty research grants from the NIH, the NSF, the Department of Defense, NASA, the American Heart Association, and other disease-oriented foundations; private industry; and individual predoctoral awards to students from organizations and sources including the American Heart Association, the NIH, NASA, the NSF, the UNCF-Merck Foundation, university teaching assistantships, university presidential fellowships, and four institutional training grants (Pharmacological Sciences NIGMS 08306, Neuropharmacology NINCDS 07254, Cell and Molecular Biology NIGMS 08109, and Lung and Cardiovascular Disease NHLBI 07050).

Students earning PhDs during the period from 1979 to 2017 as listed following and on the next page. Many of these former graduates have obtained important administrative and senior positions in academia, industry, or government or have developed well-funded research programs in academia.

- Ismail Jatai, 1981
- Dorris (Dede) Haverstick, 1982
- Stuart Dryer, 1985
- Katia Gysling, 1985
- Mark Voigt, 1985
- Alan Stephenson, 1985
- Paul Hale, 1986
- Helen McIntosh, 1986
- Roberta Secrest, 1987
- James Qualy, 1988
- William Devane, 1989
- Michelle Russell, 1989
- Anita Ciaraleglio, 1989
- Paul Zanaboni, 1990
- Eva Sorenson, 1990
- Song-ping Han, 1990

- Judith Brog, 1991
- Donald Scott, 1991
- Xiaoli Chen, 1991
- Jane Turner, 1991
- Margaret Wessels-Reiker, 1991
- Carrie Branch, 1992
- Debra DiMaggio, 1992
- John Pawloski, 1992
- Jason Viereck, 1992
- I. Lee Holt, 1993
- Lori McMahon, 1993
- J. Chris Stein, 1994
- William Weaver, 1994
- Terry Omura, 1995
- Glenn Smits, 1995
- Robert T. Dunn, 1995
- Chao Song, 1995
- Patrick Mueller, 1996
- Terese Pearman, 1996
- Mariette Baxendale, 1996
- Laura (Jansen) McCullough, 1996
- Deborah Swope, 1997
- Gui Hua Cao, 1997
- Virginia Wotring, 1997
- Lilly (Lu) Xu, 1997
- Jaeyoung Yoon, 1997
- Andrew Belt, 1997
- Sharon Bloch, 1997
- Jason Weber, 1997
- Alice Gardner, 1998
- Darren Tyson, 1998
- Weimin Hu, 1998
- Trevor Tredway, 1998
- Gonzalo Torres, 1998
- Dennis Kraichley, 1999
- Troy Baudino, 1999
- Sandra Winchester, 1999
- Robert Kirk DeLisle, 1999
- Dan Hoang, 1999
- Hobart Walling, 1999
- William Haines, 2000
- Justin Meschner, 2000
- Polly Mason, 2000
- John Swarthout, 2001
- Medora Hardy, 2001
- Richard D'Alonzo, 2001
- Susan Keenan, 2001
- Tracy Maisel Bloodgood, 2003
- Megan Taylor White, 2003
- Melanie Berken, 2003
- Derek Henley, 2004
- Lacy L. Kolo, 2004
- Reema Goel, 2004
- Jeffrey Olearczyk, 2004
- Josephine Garcia-Ferrer, 2004
- Sallee Eckler, 2004
- Adepero (Shola) Adewale, 2004
- Nathen Lents, 2004
- Chandra Coleman, 2005
- Kevin Gamber, 2005
- Jennifer (Losapio) Iverson, 2005
- Scott Ochs, 2005
- Amy Schmidt, 2005
- Donald Ruhl, 2005
- Sarah Kucenas, 2005
- Virginia Irintcheva, 2006
- Matthew Ndonwi, 2006
- Laura Jaeger, 2007
- Johnnie Moore, 2008
- Leroy Wheeler, 2008
- Laura Gorges, 2008
- Timothy Doyle, 2008
- Julie Schwartz, 2008
- Mirnela Byku, 2008
- Madeline Wong, 2008
- Angela LaMora, 2009
- Trent Moreland, 2009
- Madelyn (Stumpf) Hansen, 2009
- Tamara Bowman, 2010
- Gina Jackson, 2010
- Jesse Procknow, 2010
- Shaquira Adderley, 2010
- Sarah Burris, 2010
- Andrew Linsenbardt, 2010
- Cara Lunn, 2010
- Gina Yosten, 2010
- Meera Sridharan, 2010
- Crystal Ruth, 2011
- Michelle Erickson, 2012
- Alicia Pate, 2012
- Kelly (Thuet) Clapp, 2012
- Julie Breckenridge, 2012
- Kathryn Argue, 2012
- Jessica (Murray) Bender, 2012
- David Wilson, 2013
- Stephanie Knebel, 2013
- Kali James, 2013
- Mollisa Elrick, 2014
- Amanda (Finley) Ford, 2014
- Chloe (Bryant) Tullock, 2014
- Heather Lavezzi, 2014
- Jennifer Richards, 2014
- Holly Pope, 2014
- Julienne Tourchette, 2015
- Robert Adams, 2015
- Lauren Stein, 2016
- Carrie (Malley) Wahlman, 2016
- Rhett Reichard, 2017*
- Ryan Welch, 2017*
- Katie Stockstill, 2017*

*These students entered the program after Dr. Westfall stepped down from the chair.

Office Staff during the Westfall Era

An important component of any department is the support staff that works in the departmental office. During the time that Dr. Westfall was chair, from 1979 to 2013, the department was blessed with many outstanding individuals, and there were many dedicated people who deserve special mentions.

Joyce Jackson

Joyce was office manager of the department when Dr. René Wégria was chair. She continued in this capacity and also served as business manager when Dr. Westfall assumed the position of chair of the Department of Pharmacology. She was appointed manager of fiscal affairs when the Department of Pharmacological and Physiological Science was formed, and she served in this capacity until her retirement in 2004. Joyce was a very loyal and hardworking administrator who did a great job in helping the department run smoothly. She worked long hours, including overtime and weekends, without complaint, especially during budget and grant preparation. Her hard work and support were greatly appreciated by Dr. Westfall and the entire department. Joyce Jackson served Saint Louis University for over thirty years.

Mary Alice Kauling

Mary Alice was business manager of the Department of Physiology under Dr. Lind and was appointed manager of operations when the Department of Pharmacological and Physiological Science was formed in 1990. Mary Alice, together with Joyce Jackson, oversaw all the activities of the department and together with Dr. Westfall formed the Administrative Executive Committee. The three met on a regular basis to discuss the business and operations of the department. Upon the retirement of Joyce Jackson, Mary Alice assumed total responsibility as both business and operations manager. Like Joyce, Mary Alice was also a very loyal and hardworking administrator who put in many hours of overtime, including weekends, during budget and grant preparation time. Her loyalty and hard work were also greatly appreciated by Dr. Westfall, and she helped make his job as chair all that much easier. Mary Alice also served as business manager under Dr. Voigt's interim chairmanship and under the chairmanship of Dr. Burris until her retirement in 2014. Mary Alice served Saint Louis University for over thirty years.

Margaret "Peggy" Hubbard

Peggy Hubbard started out her career as secretary to Dr. John McGiff in the Department of Pharmacology at the University of Pennsylvania School of Medicine. She moved with Dr. McGiff when he came to Saint Louis University as head of cardiology. After a brief retirement, Peggy joined the Department of Pharmacology in the early 1980s under the chairmanship of Dr. Westfall. She continued in the office when the Department of Pharmacological and Physiological Science was formed in 1990. During much of this time, Peggy served as Dr. Westfall's personal secretary and efficiently managed all his appointments and travels. Peggy retired in 2013 after a long and distinguished career and was a dedicated member of the Saint Louis University family for over thirty years. She was a great friend to numerous medical and graduate students, who would frequently stop by her desk to chat. Peggy passed away in 2015.

Linda Russell

Linda also joined the Department of Pharmacology under Dr. Westfall's chairmanship and continued to work upon the formation of the Department of Pharmacological and Physiological Science. During most of this time she served as technology coordinator. This included maintaining the department's computers and web page and overseeing syllabus and grant preparation. Linda did an outstanding job and was a loyal member of the department and university for thirty-four years prior to her retirement in 2014. She has been indispensable in typing the many manuscript versions of this book.

Mary Busby

Mary Busby was another hard-working, dedicated, and much-loved employee who worked principally on budget issues and grant management. She first joined the Department of Pharmacological and Physiological Science in the early 2000s during Dr. Westfall's tenure as chair and continues to work under Dr. Salvemini's chairmanship. She is one of the last holdovers from the Westfall era and is much appreciated for her long hours of faithful service.

More Valuable Members of the Office Staff during the Westfall Era

- Maggie Klevorn
- Jeri-Lynn Crafton
- Rose Bradley
- Debra McGrath
- Melody Mance
- Mary Blecha
- Kathy Buseman
- Kim Stahlschmidt
- Nadine Anthony
- Mark Konsewicz
- Latonya Sayles

Research Assistants and Technicians

In addition to the office staff, who were vital to the functioning of the department, another group of individuals were extremely important to the success of the research enterprise and other aspects of the department during the Westfall era. These were the numerous research assistants and technicians that worked for the various faculty. Several of these deserve special mention because of their important roles and many years of dedicated service.

Linda Naes

Linda was a key player in the department under the chairmanships of Dr. Westfall. She spent thirty-four years of dedicated service serving as a key research assistant and lab manager of the Westfall lab. She started her research career in the Department of Surgery at the Washington University School of Medicine before moving to Saint Louis University in the VA research laboratories. She joined the Westfall lab in 1979, soon after his appointment as chair, and served faithfully until the lab closed in 2013, when Dr. Westfall stepped down from the chair. She was indispensable in setting up the lab after Dr. Westfall moved from the University of Virginia and carried out many key experiments. She soon became lab manager and in addition to conducting key experiments, she supervised all the research assistants, postdoctoral fellows, predoctoral students, summer medical students, and undergraduates for over thirty years. In addition, she was very valuable in assisting Dr. Westfall in his duties as chair. She was involved in the success of the many social activities in the department, such as graduate student graduation parties, and supervised the Christmas decorations in the department for many years. She was well known to practically everyone in the School of Medicine. Her importance was recognized when she suffered a major brain injury in an automobile accident near the end of her career. Best wishes and prayers for her recovery came from every corner of the university. Dr. Westfall owes a great deal of gratitude to Linda for her unwavering and faithful service and for the success of his research career and chairmanship.

Ismail Jatai, 1981

Dorris (Dede) Haverstick, 1982

Katia Gysling, 1985

Mark Voigt, 1985

Stuart Dryer, 1985

FIG. 31 PhD Graduates 1981-90

Alan Stephenson, 1985

Paul W. Hale Jr., 1986

Helen McIntosh, 1986

Roberta Secrest, 1987

Eva Sorenson, 1990

Paul Zanaboni, 1990

James M. Qualy, 1988

Song Ping Han, 1990

Abuta Ciaraleglio, 1989

William Devane, 1989

Margaret Wessels-Reiker, 1991 Lori McMahon, 1993 Laura (Jansen) McCullough, 1996 Glenn Smits, 1995 Deborah Swope, 1997

Gui Hua Cao, 1997 Virginia Wotring, 1997 Gonzalo Torres, 1998 Jaeyoung Yoon, 1997 Alice Gardner, 1998

Weimin Hu, 1998 Jason Weber, 1997 Andrew Belt, 1997 Sharon Bloch, 1997 Trevor Tredway, 1998 Robert Kirk DeLisle, 1999

FIG. 32 Graduate Students 1990s

FIG. 33 Graduate Students ca. 1996. Back Row: I. Lee Holt, Carrie Branch, John Pawloski, Chao Song, Patrick Mueller, Debra DiMaggio. Front Row: Jason Viereck, Robert Dunn, Terese Pearman, Xaioli Chen, Judith Brog.

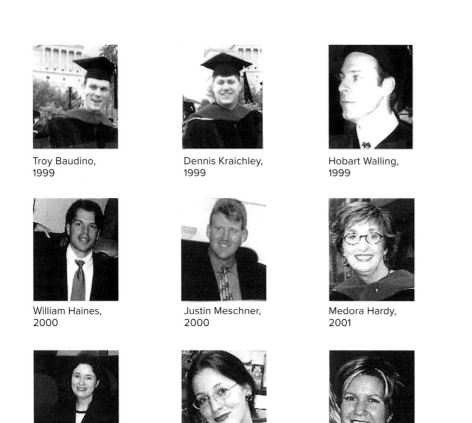

FIG. 34 Graduate Students, 1999-2004

FIG. 35. Graduate students, 2007. Front Row: Laura Jaeger, Virginia Irincheva, Sarah Kucenas, Amy Schmidt. 2nd Row: Matthew Ndonwi, Derek Henley, Jeffry Olearczyk, Melanie Berkman, Adepero (Shola) Adewale, Scott Ochs, Chandra Coleman, Charlie Travers, Donald Ruhl. 3rd Row: Jennifer (Losapio) Iverson, Gina Jackson, Reema Goel, Megan Taylor White, Kevin Gamber, Nate Lents.

FIG. 36 Graduate students, 2010. Front Row: Trent Moreland, Mirnela Byku, Leroy Wheeler, Madeline Wang, Tamara Bowman, Johnnie Moore, Angela LaMora, Gina Jackson. Back Row: Julie Schwartz, Timothy Doyle, Matthew Ndonwi, David Stroop, Madelyn (Stumpf) Hansen, Jesse Procknow, Laura Gorges.

FIG. 37 Graduate students, 2011. Front Row: Andrew Linsenbardt, Julie Breckenridge, Mirnela Byku, Sarah Burris, Julie Schwartz, Angela LaMora, Crystal Ruth, Leroy Wheeler. Back Row: Timothy Doyle, Michelle Erickson, Madeline Wang, Gina Jackson, Meera Sridharan, Shaquira Adderely, Madelyn (Stumpf) Hanson, Cara Lunn.

FIG. 38 Graduate students, 2012. Front Row: Jesse Procknow, Jessica (Murray) Bender, Meera Sridharan, Cara Lunn, Angela LaMora, Michelle Erickson, Kathy Argue, Andrew Linsenbardt, Sarah Burris. Back Row: Unknown, Stephanie Knebel, Shaquira Adderley, Tamara Bowman, Gina Jackson, Madelyn Hanson, Crystal Ruth, Gina Yosten, Julie Breckenridge.

FIG. 39 Graduate students, 2015. Front Row: David Wilson, Julienne Touchette, Katie James, Heather Lavazzi, Chloe (Bryant) Tullock, Kathryn Argue, Jessica (Bendon) Murray. Back Row: Stephanie Knebel, Amanda (Finley) Ford, Holly Pope, Alicia Pate, Kelly (Thuet) Clapp, Julie Breckenridge.

FIG 40 Graduate students, 2017. Front Row: Parag Bhatt, Heather Lavezzi, Kali James, Holly Pope, Chloe (Bryant) Tollock. 2nd Row: Jessica (Murray) Bender, David Wilson, Robert Adams, Jennifer Richards, Katie Stockstill. Back Row: Julienne Touchette, Stephanie Kneibel, Amanda (Finley) Ford, Carrie (Malroy) Wahlmann, Mollisa Elrick.

Gerald (Gerry) Wilken

Gerry started his research career in the department in Dr. Allyn Howlett's laboratory and was an important contributor to her research success in defining the mechanism of action of tetrahydrocannabinol (cannabinoid) drugs. Following Allyn's move to North Carolina, Gerry worked in Heather Macarthur's lab, with joint duties in Dr. Westfall's lab. He continued with Dr. Macarthur when Dr. Westfall closed his laboratory. Gerry was not only a very valuable research assistant but also a very loyal and faithful participant in all departmental activities. He could be counted upon to assist in numerous ways and was a fixture in the department for over thirty-five years.

Jo Schrewis

Jo was another person who was a prominent fixture in the department for over thirty years. She originally served in Dr. Andrew Lonigro's lab and later worked in the laboratories of Dr. Sprague and Dr. Stephenson. She was very accomplished and contributed a great deal to the success of these three faculty members.

Other research assistants that served faithfully in the Department of Pharmacology and Physiology for many years include the following:

Over 15 Years of Service
- Lloyd Allard (Beinfeld lab)
- Kathleen Wolf (Chiappinelli lab)
- Jason Papke (Harkins lab)

Over 30 Years of Service
- David Achilleus (Ellsworth lab)
- Linda Cox (M. Ruh lab)
- Elizabeth Bowles (Sprague lab)
- Chun-Lian Yang (Westfall lab)
- Lillian Vickery (Westfall lab)

TABLE 35 Summary of Accomplishments for the Department of Pharmacology (1979–90) and the Department of Pharmacological and Physiological Science (1990–2012)

Extramural Support Generated	$86,994,537
Peer-Reviewed Papers Published	over 1500
PhDs Awarded	126
Training Grant in Neuropharmacology (NINCDS) NS16215 (1985–95)	$1,232,245
Training Grant in Pharmacological Sciences (NIGMS) GM06306 (1990–2015)	$2,964,629
Membership on Permanent NIH Study Sections	37
Participation on Special NIH Emphasis Panels	284
Participation on Non-NIH Review Panels	271
Speaking Invitations: National/International Meetings	315
Speaking Invitations: Seminars at Other Institutions	496
Membership on Editorial Boards	46
Participation in Medical School, Graduate School, and Service Courses	40
Course Directorships	20
Membership on National/International Committees	241
Membership on University Committees	130
Former Members Who Assumed Leadership Roles in Academic Departments	8

TABLE 36 Former and Current Members of the Faculty Who Assumed Leadership Roles in Academia

Name	Leadership Position
Dr. Vincent Chiappinelli	Chairman, Department of Pharmacology and Physiology; Dean of Research, George Washington University School of Medicine
Dr. Nicola Partridge	Chair, Department of Physiology, Robert Wood Johnson School of Medicine, University of Medicine and Dentistry of New Jersey
	Chair, Department of Basic Sciences, New York University School of Dentistry
Dr. Francis White	Chair, Department of Pharmacology, Rosiland Franklin School of Medicine (Chicago) (Deceased)
Dr. Michael Meldrum	Chair, Department of Pharmacodynamics, University of Florida College of Pharmacy
Dr. Allyn Howlett	Director, NIDA Minority Training Program, North Carolina Central University
Dr. Mark Voigt	Vice Chair and Interim Chair, Department of Pharmacology and Physiology; Director, Center for Neuroscience, Saint Louis University School of Medicine; Associate Dean of Research, Doisy College of Health Sciences, Saint Louis University
Dr. Amy Harkins	Chair, Department of Health Sciences and Informatics, Doisy College of Health Sciences, Saint Louis University
Dr. Daniela Salvemini	Vice Chair-Research, Department of Pharmacology and Physiology; Director, Henry and Amelia Nasrallah Center for Neuroscience, Saint Louis University School of Medicine; Interim Chair, Department of Pharmacology and Physiology

DEPARTMENT OF PHARMACOLOGY and PHYSIOLOGY: The THOMAS P. BURRIS ERA and the MARK VOIGT ERA

2013–2018 and 2018–2019

Biographical Sketch of Thomas Burris, PhD

Dr. Burris was born on September 15, 1968, and grew up on the east side of metropolitan St. Louis, where he went to Cahokia High School. He earned a bachelor of arts in chemistry (magna cum laude) from Southern Illinois University Edwardsville in 1989 and followed this with a doctor of philosophy in molecular biophysics (biochemistry molecular and cellular biology track) at the Institute of Molecular Biophysics at Florida State University in 1992. His advisor was Dr. Marc E. Freeman. Dr. Burris was a postdoctoral fellow in molecular endocrinology in the Department of Cell Biology at the Baylor College of Medicine from 1993 to 1994, working in the laboratory of Dr. Bert W. O'Malley. This was followed by a second postdoctoral fellowship in human genetics in the Division of Genetics in the Department of Pediatrics at the Molecular Biology Institute of the UCLA School of Medicine, working in the laboratory of Dr. Edward R. B. McCabe from 1994 to 1995. He was a scientist in 1996 and 1997 and then a senior scientist from 1997 to 1999 in drug discovery at Johnson & Johnson Pharmaceutical Research and Development. Dr. Burris then assumed a variety of positions in discovery biology at Lilly Research Laboratories, including research scientist from 1999 to 2003, head of nuclear receptor biology research from 2001 to 2006, senior research scientist from 2003 to 2004, research advisor from 2004 to 2005, and senior research advisor from 2004 to 2006. He was appointed professor (with tenure) at the Pennington Biomedical Research Center at Louisiana State University from 2006 to 2008. He was appointed professor (with tenure) in the Department of Molecular Therapeutics at the Scripps Research Institute, a position he held until 2013. He was also a professor in the Department of Metabolism and Aging at Scripps those same years, a member of the graduate faculty at the Kellogg School of Science and Technology at Scripps from 2009 to 2013, and director of the Center for Diabetes and Metabolic Diseases at Scripps from 2011 to 2013. In July 2013 Dr. Burris was appointed the second William Beaumont Professor and Chair of the Department of Pharmacological and Physiological Science, which was renamed the Department of Pharmacology and Physiology, at the Saint Louis University School of Medicine. He was also a member of the Alvin J. Siteman Cancer Center's Solid Tumor Therapeutics Program at the Washington University School of Medicine. In addition, he has been an adjunct faculty member at several institutions including the Department of Pharmacology at the School of Medicine and Dentistry of New Jersey, Department of Pharmacology and Toxicology at the Indiana University School of Medicine, and

FIG. 41 Dr. Thomas Burris

the Department of Biological Sciences at Louisiana State University.

Dr. Burris's research has been in the area of chemical biology, drug development, and nuclear receptors and their action and pharmacology, particularly in the area of chemical biology of orphan nuclear receptors. He has pioneered the use of REV-ERB and ROR ligands for the treatment of a wide variety of disorders, including anxiety, circadian rhythm disturbances, metabolic disorders, steatohepatitis, muscular dystrophy, and cancer. He has published over 150 peer-reviewed articles and two books and has four patents with nine patents pending. He has been awarded numerous grants as principal investigator from the NIH, the Department of Defense, the American Heart Association, and private industry. He was also co-principal investigator of a T32 training grant from NIGMS. Dr. Burris currently serves on six journal editorial boards and two scientific advisory boards for pharmaceutical companies. He has chaired nine major international conferences and has served on fourteen NIH review panels, eleven Department of Defense panels, and numerous other government and research foundations. He is a consultant to thirteen pharmaceutical companies. He is a fellow of the American Heart Association, the American Association for the Advancement of Science, and the St. Louis Academy of Science. He has also been commissioned a Kentucky Colonel by Steven Beshear, Governor of Kentucky. He has been an invited speaker to fifty national or international conferences and symposia and has given seventy-two invited lectures and seminars worldwide. To date he has mentored seven PhD students, fourteen postdoctoral fellows, and three undergraduate students and has served on nine PhD dissertation committees. He also participates in departmental teaching programs to medical and graduate students.

Because of the uncertainties of the planned reorganization and restructuring of the School of Medicine at Saint Louis University proposed by Dean Kevin Behrns (especially the future of basic sciences), Dr. Burris felt that his research program in chemical biology and drug discovery would prosper better at another institution. He therefore resigned his position as William Beaumont Professor and Chair of the Department of Pharmacology and Physiology in February 2018. He accepted a position as professor and special assistant to the president of the St. Louis College of Pharmacy with joint appointments in the Departments of Anesthesiology and Genetics at the Washington University in St. Louis School of Medicine. On February 5, 2018, Dr. Mark Voigt assumed the role of interim chair of the Department of Pharmacology and Physiology at the Saint Louis University School of Medicine.

Recruitment of Dr. Thomas Burris

Following the announcement by Dr. Thomas Westfall that he was voluntarily stepping down from the chair, a search committee was formed by Dean Alderson. He appointed Dr. Enrico Di Cera, chair of the Department of Biochemistry and Molecular Biology, as chair of the search committee, with instruction to conduct a national search to find a replacement for Dr. Westfall. Dr. Westfall relinquished the chair in January 2013; Dr. Mark Voigt was appointed interim chair and served until July 1, 2013. The search committee identified several outstanding candidates, and four of these were invited to St. Louis for interviews. One of those four was Dr. Thomas Burris, who ultimately was the successful candidate. Dr. Burris was interested in becoming a chair and initially went to the Scripps Institute with the idea of chairing the Department of Metabolism and Aging, but he decided not to accept the position at that institution. He was a candidate for the chair of Biochemistry at the University of Louisville School of Medicine but withdrew from consideration. Dr. Burris was interested in bringing drug discovery and chemical biology to academia and was aware that Saint Louis University had hired a number of scientists from Pfizer, a pharmaceutical company, to put together a program to develop drugs for rare diseases. He saw the advertisement for the chair of the Department of Pharmacological and Physiological Science and submitted an application. He was soon contacted by Dr. Di Cera and was invited by the search committee to visit Saint Louis University as a candidate for the chair's position. Following three visits, Dr. Burris was offered the position of William Beaumont Professor and Chair of the Department of Pharmacological and Physiological Science. Following discussions, which included a very competitive setup package along with funds to purchase several major items of equipment, Dr. Burris agreed to accept the offer. He was appointed July 1, 2013, and served until February 2018.

Dr. Burris's Mission and Vision for the Department of Pharmacology and Physiology

Dr. Burris had several goals for the department, both in education and training and in research.

Education and Training

1. Maintain a culture of excellence in educating and training medical students and graduate students as established in the department in the past.
2. Ensure renewal of the NIH T32 training grant in Pharmacological Sciences as established by Dr. Westfall and the departme\nt since 1990 (twenty-five years of continual funding).
3. Require trainees to develop independent grant applications and reward success.
4. Ensure that PhD trainees are prepared for diverse scientific careers.
5. Enhance funding to entice high-quality students to train in the department at Saint Louis University.

Research

Dr. Burris's plan was to shift the focus of the research direction of the department to one that emphasized drug discovery and development. This theme was being developed at several institutions nationally due to decreased interest in the pharmaceutical industry in maintaining basic research laboratories. In order to be successful in this area of research Dr. Burris maintained that faculty must act as entrepreneurs and follow the NIH and the interest in translational research. He thought it was important to continue developing collaborative and interdisciplinary team science and develop an innovative translational focus. Moreover, it was important to enhance current strengths of the department with new capabilities.

In order to accomplish this, he planned to create significant chemical biology capabilities within the department and recruited five new faculty, two of them medicinal chemists. He also began to establish infrastructure for chemical biology by interfacing with the Center for World Health and Medicine until it was dissolved by Dean Behrns; he promoted collaboration between chemists and current faculty to progress into development of novel chemical probes for translational research and planned to do additional faculty recruiting focused on the ability to utilize chemical biology approaches. Dr. Burris also tried to ensure that necessary core research infrastructure was available to the department. Although well underway, these plans never completely materialized due to his resignation in 2018.

Departmental Administration and Committee Structure

Dr. Thomas Burris was assisted by an office staff that was reduced in size from that in the Westfall era by approximately three to six positions, depending on the year, one of whom was business manager Stacy Smith. Dr. Burris was also assisted in administration of the department by two vice chairs, Dr. Mark Voigt and Dr. Daniela Salvemini. One of these represented Dr. Burris at key meetings in his absence and advised him in all important departmental decisions. There was also an elaborate committee structure consisting of the following committees (as of January 25, 2018):

Seminar Committee
- Andrew Butler, Chair
- Gina Yosten
- John Walker
- Jinsong Zhang

Graduate Education Steering Committee
- Terry Egan, Chair
- Andrew Lechner
- Gina Yosten
- Jinsong Zhang
- Heather Macarthur
- Daniela Salvemini
- Mark Voigt

Undergraduate Education Committee
- John Chrivia, Chair
- Andrew Lechner
- Joe Baldassare
- John Walker

Medical Education Committee
- Andrew Lechner, Chair
- Heather Macarthur
- Rick Samson
- John Chrivia
- Mark Knuepfer

Postgraduate Education Committee
- Daniela Salvemini, Chair
- Terry Egan
- Colin Flaveny
- Andrew Lechner
- Mark Knuepfer

Tenure/Promotion (Mentoring) Committee
- Rick Samson, Chair
- Mark Voigt
- Daniela Salvemini
- Terry Egan
- Andrew Lechner

Communication (Website, Facebook, etc.) Committee
- Jinsong Zhang, Chair
- Andrew Butler

Preliminary Exam Standing Committee
- Heather Macarthur, Chair
- Andrew Lechner
- Jinsong Zhang
- John Walker
- Michael Ariel

Departmental Executive Committee (Policies) Committee
- Thomas Burris, Chair
- Andrew Lechner
- Mark Voigt
- Daniela Salvemini
- Willis Samson
- Stacy Smith

Journal Club Committee
- Daniel Zahm, Chair
- Terry Egan
- Willis Samson
- Colin Flaveny

Departmental Historian
- Thomas Westfall

Faculty Meetings and Retreats

Faculty meetings were held monthly and retreats held biannually. Table 37 shows the agenda for a faculty meeting held on January 27, 2017, which was representative of all meetings.

Typical retreats consisted of an update on the state of the department, school, and university by Dr. Burris and a business meeting in which reports were provided by chairs of the various standing committees. This part of the retreats was for faculty only and was generally followed by lunch, to which the entire department was invited. In the afternoon, there were oral presentations by graduate students competing for the Thomas Westfall Fellowship Award. The winner was awarded a $2,500 travel award and a handsome crystal (pictured in Fig. 42). A panel of judges consisting of faculty who did not have students competing chose the winners, named in Table 38.

Medical Education Activities

During the Burris era, the department's participation in the medical school's teaching program remained

TABLE 37 Pharmacology and Physiology Departmental Faculty Meeting Agenda, January 27, 2017

I. Announcements
   a. Staff/Faculty member update
   b. New SOM leadership
   c. Faculty Senate News - Knuepfer
II. Updates
   a. Committee Updates
      i. Seminar Committee
      ii. Graduate Education
         1. Student Reviews of Pharmacology & Physiology Courses - Egan
         2. PR minority recruitment - Knuepfer
      iii. Undergraduate Education
      iv. Medical Education
      v. Postgraduate Education
      vi. Communication
      vii. Preliminary Exam
      viii. Journal Club
      ix. Tenure/Promotion/Mentoring
      x. Executive/Policy
   b. Funding and Ranking update
   c. Departmental Retreat - date selection
III. New Business
   a. May 18 Pre-commencement Ceremony
   b. Walk-in items

TABLE 38 Winners of the Thomas C. Westfall Fellowship Award

Student	Mentor
Lauren M. Stein	Yosten and Samson
Rhett Reichard	Zahm
Carrie (Maloy) Wahlman	Salvemini
Nickolas Steinauer	Zhang
Katie Carpentier	Burris
Kathryn Breden McInerny	Salvemini

essentially the same as during the second part of the Westfall era. Dr. Andrew Lechner remained course director for the hematology/respiratory module and Dr. Heather Macarthur assumed the role of course master for the principles of pharmacology module. Faculty who participated in the various medical student teaching activities included Dr. Ariel, Dr. Baldassare, Dr. Butler, Dr. Egan, Dr. Flaveny, Dr. Knuepfer, Dr. Lechner, Dr. Macarthur, Dr. Neckameyer, Dr. Samson, Dr. Voigt, Dr. Walker, Dr. Yosten, Dr. Zahm, and Dr. Zhang. Table 39 shows the participation of the department's faculty in the following medical school modules: cell and molecular biology, principles of pharmacology, cardiovascular, hematology/respiratory, neuroscience, behavioral medicine and health, and endocrine/reproduction. They gave similar lectures or otherwise participated in a similar fashion as described earlier in Chapter Ten.

Table 40 shows the participation of the department's faculty in administrative activities of the school of medicine.

FIG. 42 The Thomas Westfall Fellowship Award

TABLE 39 Medical Student Teaching (2016–17)

Faculty Member	Phase 1 Foundational Courses	Phase 2 System Modules	MD Elective Offered?	Phase 3 Courses
Ariel		Neurosciences		
Baldassare	Cell & Molec. Biology; Prin. Pharmacology			
Burris				
Butler		Cardiovascular		
Chrivia	Prin. Pharmacology			
Cox				
Egan	Cell & Molec. Biology			
Flaveny	Cell & Molec. Biology			
Harkins				
Knuepfer	Prin. Pharmacology	Neurosciences	Yes	
Lechner		Hematology; Respiratory		Surgery Capstone
Macarthur	Prin. Pharmacology	Neurosciences; Cardiovascular		
Salvemini				
Samson	Cell & Molec. Biology	Neurosciences; Endocrine	Yes	
Sprague				
Voigt		Behavior Medicine & Health; Neurosciences; Endocrine		
Walker	Cell & Molec. Biology; Prin. Pharmacology			
Yosten		Hematology; Endocrine	Yes	
Zahm		Neurosciences		
Zhang	Cell & Molec. Biology			

TABLE 40 Medical Student Administration (2016–17)

Faculty Member	MD Residents/MD Fellows?	MD-related Committees	Other MD-related
Ariel			
Baldassare			
Burris			
Butler			
Chrivia			
Cox			
Egan			
Flaveny			
Harkins		Student Progress & Promotion	LCME Site Visit Team
Knuepfer			LCME Site Visit Team
Lechner	IM Res.; Pulm/Crit. Care Fellows	Admissions; Phase 2 oversight	LCME Site Visit Team
Macarthur		Curriculum Oversight	LCME Site Visit Team
Salvemini			
Samson		Phase 2 oversight	LCME Site Visit Team
Sprague			
Voigt			
Walker			
Yosten			
Zahm			
Zhang	Hem/Onc Fellows		

TABLE 41 Principles of Pharmacology Schedule, 2018

Date and Time	Topic	Lecturer	Location
Week 1			
January 2 Tuesday			
9-9:30	Introduction to Course	Dr. Macarthur	EU Aud
9:30-12:00	Topic 1: Drug Absorption/Distribution	Dr. Baldassare	EU Aud
January 4 Thursday			
9-10:30	Topic 2: Biotransformation	Dr. Walker	EU Aud
10:30-11:30	Topic 3: Pharmacokinetics	Dr. Baldassare	EU Aud
11:30-12:00	Clinical Correlation	Dr. Sprague	EU Aud
1:00-3:00	Topic 4: Pharmacokinetic Problems	Dr. Baldassare	EU Aud
January 5 Friday			
9-10:30	Topic 5: Mechanism of Drug Action	Dr. Baldassare	EU Aud
10:30-12:00	Topic 6: Factors Modifying Drug Action	Dr. Knuepfer	EU Aud
1-2:00	Topic 7: Geropharmacology	Dr. Little	EU Aud
Week 2			
January 8 Monday			
9-9:50	Quiz 1		
10-11:00	Topic 8: Autonomic & Somatic Nervous System	Dr. Macarthur	EU Aud
11-12:00	Topic 9: Adrenergic Transmission	Dr. Macarthur	EU Aud
1-2:00	Topic 10: Adrenergic Drugs	Dr. Macarthur	EU Aud
2-2:30	Clinical Correlation 2	Dr. Sprague	EU Aud

TABLE 41 Principles of Pharmacology Schedule, 2018, *cont'd*

Date and Time	Topic	Lecturer	Location
Week 2, *cont'd*			
January 9 Tuesday			
10-11:00	Topic 11: Cholinergic Transmission	Dr. Knuepfer	EU Aud
11-12:00	Topic 12: Cholinergic Drugs	Dr. Knuepfer	EU Aud
January 10 Wednesday			
9-10:15	Simulation Labs Group 1	Drs. Macarthur and Knuepfer	LRC 101
10:30-12:00	Simulation Labs Group 2	Drs. Macarthur and Knuepfer	LRC 101
January 11 Thursday			
9-10:15	Simulation Labs Groups 3 & 4	Drs. Macarthur and Knuepfer	LRC 101
10:30-12:00	Simulation Labs Groups 5 & 6	Drs. Macarthur and Knuepfer	LRC 101
January 12 Friday			
9-10:15	Simulation Labs Groups 7 & 8	Drs. Macarthur and Knuepfer	LRC 101
10:30-12:00	Simulation Labs Groups 9 & 10	Drs. Macarthur and Knuepfer	LRC 101
1-2:15	Simulation Labs Groups 11 & 12	Drs. Macarthur and Knuepfer	LRC 101
3:30-4:00	Simulation Labs Groups 13 & 14	Drs. Macarthur and Knuepfer	LRC 101
Week 3			
January 16 Tuesday			
9-9:50	Quiz 2		EU Aud
10-11:00	Topic 13: Introduction to Antimicrobes	Dr. Walker	EU Aud
11-12:00	Topic 14: Sulfonamides & Quinolones	Dr. Chrivia	EU Aud
January 18 Thursday			
9-10:00	Topic 15: b-Lactams	Dr. Walker	EU Aud
10-11:00	Topic 16: Aminoglycosides & Tetracyclines	Dr. Knuepfer	EU Aud
11-12:00	Topic 17: Miss Antibiotics	Dr. Chrivia	EU Aud
January 19 Friday			
9-10:00	Topic 18: Antimyobacterium	Dr. Chrivia	EU Aud
10-11:00	Topic 19: Antifungals	Dr. Knuepfer	EU Aud
11-12:00	Topic 20: Antiprotazoals	Dr. Chrivia	EU Aud
Week 4			
January 22 Monday			
9-10:30	Topic 21: Antivirals	Dr. Wold	EU Aud
10:30-11:30	Topic 22: Antihelminths	Dr. Chrivia	EU Aud
11:30-12:00	Clinical Correlation 3	Dr. Sprague	
1-3:00	Topic 23: Cancer Chemotherapy	Dr. Chrivia	EU Aud
January 23 Tuesday			
10-12:00	Topic 24: Cancer Chemotherapy	Dr. Chrivia	EU Aud
January 24 Wednesday			
9-9:50	Quiz 3		EU Aud
January 25 Thursday			
	Study Day		
January 26 Friday			
9-11:00	NBME Exam		EU Aud

Principles of Pharmacology

The objectives of the Principles of Pharmacology Module and the structure of the course remained essentially the same as described in Chapter Ten. An updated schedule and lecture topics for the 2018 course are shown in Table 41.

Graduate Program and Educational Activities

The graduate program that was developed during the Westfall era has been maintained under the leadership of Dr. Burris, Dr. Voigt, and now Dr. Salvemini. The numbers of students entering the program and graduating are also similar.

Graduate Courses

The formal coursework for PhD students under Dr. Burris and Dr. Voigt remained quite similar to that under Dr. Westfall. However, the second-year course (PPY511–514) has been continuously tweaked to better reflect the current faculty and objectives of the department. PPY511, 512, and 513 were presented as separate modules under the topics of pharmacology, neurobiology, systems physiology, endocrinology and metabolism, and integrative physiology and pharmacology. Placing these topics into separate modules carrying different credit hours allowed for students in other programs to participate in individual modules without having to take the entire course. When Dr. Westfall retired from the chair, he also relinquished being program director of the T32 training grant in Pharmacological Sciences, a position he held from 1990 to 2015. Leadership of the grant was divided among three co-program directors, Tom Burris, Mark Voigt, and Terry Egan, each of whom had well-defined duties. In 2015 the training grant was submitted for the next five-year cycle. The first submission was not funded, but the application was resubmitted, resulting in a site visit. This time the T32 was approved for an additional five years (2016–21), with four slots per year. The renewed training grant was somewhat refocused, with chemical biology and drug discovery added as a theme. Following Dr. Burris's resignation as chair in 2018, John Walker replaced him as co-program director.

Table 43 shows a representative schedule for the PPY511, 512, and 513 courses for the 2015–16 semesters.

Grant-Writing Course

As a result of the critique of the resubmission of the T32 training grant in Pharmacological Sciences, the

---

Table 42

Constructing Grant Applications in the Pharmacological Sciences
PPY514, 1 Credit Hour   Spring 2017
COURSE DIRECTOR: Gina L. C. Yosten, PhD

LEARNING OBJECTIVES

Gain a basic understanding of the types of funding mechanisms offered by the major grant funding agencies and the required application elements needed to apply for those awards.

Be able to develop an outline for a research project, and succinctly and effectively communicate the goals, hypotheses, and experimental design of the study.

Understand the fundamentals of constructing an NIH-style grant application.

Understand how a study section operates and how to review a grant.

COURSE STRUCTURE

The class will meet twice per week for 15 weeks in 2-hour class periods, and will consist of a mixture of didactic lectures, mentoring sessions, and dedicated writing time. Each student will be required to write an NIH-style R01 grant application, which can be used as the written portion of their qualifying examination. The grant proposal may be related or unrelated to the student's project, but must incorporate concepts in the Pharmacological or Physiological Sciences.

COURSE ELEMENTS

1. Lectures: Pharm/Phys faculty will provide lectures on funding mechanisms through major grant funding agencies and on the fundamentals of writing an effective grant proposal.
2. Dedicated Writing Time: Students will have class periods designated as dedicated writing time for constructing specific portions of their proposals. Each student's mentoring team will be available during this time to answer any questions.
3. Mentoring Sessions: Each student will choose at least 2 Pharmacology faculty to form their mentoring team, including their dissertation advisor and up to 2 additional Pharmacology faculty. The mentoring team will meet with the student to develop outlines of the proposal sections, and to provide feedback as the proposal develops. *Mentoring teams must be established and reported to the Course Director by April 18, 2017*
4. Mock Study Section: At the end of the course, the students will engage in a mock study section. Each student will be required to review one grant application (supplied by the faculty), and present their review during the mock study section. Guidelines on the mock study section are at the end of the syllabus.

EVALUATION CRITERIA

Students will be graded on class participation, writing assignments, and the final oral presentation, as detailed below:

1. Class Participation (10%, 100 points possible): Lecturers will assign a letter grade for each student on lecture days according to their participation in the discussion.
2. Grant Proposal (80%, 800 points possible): Each student will be required to construct an NIH-style R01 grant proposal that must include Specific Aims (1 page), the Research Proposal (12 pages total for Significance, Innovation, and Research Design sections), and Literature Cited (no page limits). A grading committee, independent of the mentoring committees and consisting of 2-3 faculty, will assign a letter grade to the proposal. The proposals will be graded on completeness, clarity, and soundness of the research plan. The proposal is due one week following the mock study section.
3. Mock Study Section (10%, 100 points possible): Each student will present their critique of their assigned grant proposal, using the NIH critique template provided. Grading will be based on participation and engagement in the mock study section (evaluated by the faculty running the mock study section). See required preparation at the end of the syllabus.

GRADING SCALE:
900–1000 points A
850–900 points A–
800–849 points B
750–799 points B–
700–750 points Remediation permitted (see following)
Below 700 points F, no remediation is permitted in the same semester. The student must retake the course at the next available opportunity.

Remediation: Any student receiving a grade of 75% or below will be granted one opportunity to remediate. Remediation will include 1 month to correct any deficits in the grant proposal, followed by re-evaluation by the student's mentoring team, the grading committee, and the course director. If remediation is required to attain a passing grade, then the highest possible grade awarded will be a B (3.0).

TABLE 42 Course Schedule: Grant-Writing Course

Date	Day	Topic	Objective	Lecturer
April 21	Friday	Lecture: Overview of Funding Agencies and Programs		Burris
April 24	Monday	Lecture: Constructing the Specific Aims	Homework: Develop Rough Draft of Specific Aims Outline	Egan
April 25	Tuesday	Mentoring Session: Develop Outline of Specific Aims	Homework: Develop Specific Aims Outline	Mentoring Teams
April 27	Thursday	Dedicated Writing Time: Specific Aims	Write Specific Aims Based on the Outline Developed with the Mentors	Mentoring Teams Available
May 2	Tuesday	Mentoring Session: Specific Aims Revision	Discuss and Revise the Specific Aims Page	Mentoring Teams
May 4	Thursday	Dedicated Writing Time: Specific Aims	Revise Specific Aims	Mentoring Teams Available
May 9	Tuesday	Mentoring Session: Specific Aims Revision	Discuss and Revise the Specific Aims Page	Mentoring Teams
May 11	Thursday	Dedicated Writing Time: Specific Aims	Final Revisions of Specific Aims	Mentoring Teams Available
May 16	Tuesday	Lecture: Developing the Significance Section	Homework: Develop Rough Outline of Significance Section	Voigt
May 18	Thursday	Mentoring Session: Develop Outline of Significance section	Develop Outline of Significance Section	Mentoring Teams
May 23	Tuesday	Dedicated Writing Time: Significance	Write Significance Section Based on the Outline Developed with the Mentors	Mentoring Teams Available
May 25	Thursday	Mentoring Session: Revise Significance Section		Mentoring Teams
May 30	Tuesday	Dedicated Writing Time: Significance	Revise Significance Section Based on the Discussion with the Mentors	Mentoring Teams Available
June 1	Thursday	Lecture: Identifying the Innovative Elements of the Proposal	Homework: Create Rough Outline of Innovation Section	Zahm
June 6	Tuesday	Mentoring Session: Develop outline of Innovation		Mentoring Teams
June 8	Thursday	Dedicated Writing Time: Innovation Section	Write Innovation Section Based on the Outline Developed with the Mentoring Team	Mentoring Teams Available
June 13	Tuesday	Mentoring Session: Discussion/Revision of Innovation Section		Mentoring Teams
June 14	Wednesday	Lecture: Constructing an Effective Research Design Section	Homework: Develop Rough Draft of the Research Plan	Samson
June 20	Tuesday	Dedicated Writing Time: Revise Innovation Section and Outline Research Plan		
June 22	Thursday	Mentoring Session: Develop Outline of the Research Design Section		Mentoring Teams
June 27	Tuesday	Dedicated Writing Day: Research Design (Part 1)	Write First Half of the Research Plan Based on the Outline Developed with the Mentoring Team	Mentoring Teams Available

TABLE 42 Course Schedule: Grant-Writing Course, *cont'd*

Date	Day	Topic	Objective	Lecturer
June 29	Thursday	Mentoring Session: Discussion/Revision of Part 1 of the Research Design Section		Mentoring Teams
July 5	Wednesday	Dedicated Writing Time: Research Design (Part 1)	Revise First Half of the Research Plan Based on the Discussion with the Mentoring Team	Mentoring Teams Available
July 7	Friday	Mentoring Session: Discussion/Revision of Research Design, Part 1; Review Outline for Part 2		Mentoring Teams
July 11	Tuesday	Dedicated Writing Time: Research Design (Part 2)	Write Second Half of the Research Plan Based on the Outline Developed with the Mentoring Team	Mentoring Teams Available
July 13	Thursday	Mentoring Session: Discussion/Revision of Part 2 of the Research Design Section		Mentoring Teams
July 18	Tuesday	Dedicated Writing Time: Research Design (Part 2)	Revise Second Half of the Research Plan Based on the Discussion with the Mentoring Team	Mentoring Teams Available
July 20	Thursday	Mentoring Session: Discussion/Revision of Research Design	Finish Final Revisions of Proposal	Mentoring Teams
July 25	Tuesday	Mentoring Session: Final Discussion of Research Proposals	Homework: Make Final Revisions to Proposal	Mentoring Teams
July 28	Thursday	Mock Study Section		All Lecturers and Mentoring Teams

Mock Study Section Preparation

Each student will be assigned one grant proposal to review (provided by the faculty), and should evaluate the proposal as if they are "Reviewer 1." Each grant proposal will be evaluated by 2 students and 1 faculty member. Although the faculty reviewer will provide feedback on the students' reviews during the mock study section, the grade for the mock study section will be based solely on the students' participation (engagement in the mock study section, including asking questions and discussing relevant points of the grant proposals). Grant assignments and mock study section schedule will be distributed by the end of June. Prior to the mock study section, students should complete the following tasks:

1. Read the NIH guidelines on preparing a critique (1 page) and NIH Reviewer Orientation (attached). The NIH Reviewer Orientation is provided as a PDF (7 pages), or can be found at: https://grants.nih.gov/grants/peer/guidelines_general/reviewer_orientation.pdf
2. Watch the NIH Mock Study Section video (17 minutes), found at the link below: https://www.youtube.com/watch?v=IzBhKeR6VIE
3. Prepare the critique of your assigned grant proposal using the NIH review template. You will use this template to present your critique during the Mock Study Section.

TABLE 43 PPY511, 512, and 513 (2015–16)

Subtopic	Lectures	Dates	Course Title
	Each line represents a 2-hour lecture or other event		
Pharmacology			PPY511
Pharmacology	Introduction to Pharmacology	Monday, August 24, 2015	1 Credit Hour
	Binding Theory	Tuesday, August 25, 2015	
	Efficacy and Potency	Thursday August 27, 2015	
	Partial Agonists and Antagonists	Friday, August 28, 2015	
	Allosteric Modulators	Monday, August 31, 2015	
	Quantitative Pharmacology: Technology	Tuesday, September 1, 2015	
	Quantitative Pharmacology: Statistical Tools	Thursday, September 3, 2015	
	Problem-Based Practice and Review	Friday, September 4, 2015	
	LABOR DAY	Monday, September 7, 2015	
	Drug Metabolism	Tuesday, September 8, 2015	
	Pharmacokinetics	Thursday, September 10, 2015	
Drug Development	Basic Principles of Medicinal Chemistry	Friday, September 11, 2015	
	Structure-Activity Relationships	Monday, September 14, 2015	
	Problem-Based Practice and Review	Tuesday, September 15, 2015	
	Clinical Trials and Drug Approval Process	Thursday, September 17, 2015	
	MODULE 1 EXAM	Friday, September 18, 2015	
	Module 1 Hypothesis and Experimental Design	Monday, September 21, 2015	
Module 2: Neurobiology			PPY512
Neuroanatomy Overview	CNS, PNS and ANS; hypothalamus and limbic system	Tuesday, September 22, 2015	3 Credit Hours
Neurodevelopment	Genesis of cells and establishment of polarity	Thursday, September 24, 2015	
	Axon guidance and synaptogenesis	Friday, September 25, 2015	
Neurophysiology	Synapses, volume transmission; excitatory neurotransmission	Monday, September 28, 2015	
	Inhibitory neurotransmission; metabotropic receptor signaling	Tuesday, September 29, 2015	
Sensory systems	Sensory transduction	Thursday, October 1, 2015	
	Sensory pathways	Friday, October 2, 2015	
	Pain: nociceptive circuits - peripheral, spinal, and supraspinal pathways	Monday, October 5, 2015	
Motor systems	Corticospinal and corticobulbar systems, cerebellum - motor control	Tuesday, October 6, 2015	
	Basal ganglia and the emotional motor system	Thursday, October 8, 2015	
	MODULE 2 EXAM 1	Friday, October 9, 2015	
Neuropharmacology	Overview of neuropharmacology	Monday, October 12, 2015	
	Anticonvulsants, sedatives, and hypnotics	Tuesday, October 13, 2015	
	Monoamines	Thursday, October 15, 2015	
	Serotonin and cholinergics	Friday, October 16, 2015	
	FALL BREAK	October 19-20, 2015	

Subtopic	Lectures	Dates	Course Title
Glia	Glia: homeostasis and signal transduction	Thursday, October 22, 2015	PPY512
Neurodegeneration	Neurodegeneration	Friday, October 23, 2015	3 Credit Hours
Integrative neurobiology	Neuroplasticity	Monday, October 26, 2015	
	Reward and addiction	Tuesday, October 27, 2015	
	Memory and learning	Thursday, October 29, 2015	
	Pain: control of nociception - therapeutics	Friday, October 30, 2015	
	MODULE 2	Tuesday, November 3, 2015	
	Special Topic - Molecular intervention in brain: viral gene transfer, optogenetics	Thursday, November 5, 2015	
Module 3: Systems Physiology			
Cardiovascular	Homeostasis - Claude Bernard	Friday, November 6, 2015	
	Autonomic Nervous System Physiology	Monday, November 9, 2015	
	Autonomic Nervous System Pharmacology (Cholinergics/Adrenergics)	Tuesday, November 10, 2015	
	Regulation of Arterial Pressure	Thursday, November 12, 2015	
	Systemic Circuitry and Hemodynamics	Friday, November 13, 2015	
	Local Control of Perfusion and Vascular Mediators	Monday, November 16, 2015	
	Cardiac Physiology	Tuesday, November 17, 2015	
	Cardiac Pharmacology	Thursday, November 19, 2015	
	Platelets and Vascular Pathology	Friday, November 20, 2015	
	MODULE 3 EXAM 1	Monday, November 23, 2015	
	THANKSGIVING WEEK	November 23-27, 2015	
Kidney	Renal Hemodynamics	Monday, November 30, 2015	
	Glomerular Filtration	Tuesday, December 1, 2015	
	Renal Tubular Function	Thursday, December 3, 2015	
	Renal Mechanisms of Acid - Base Balance	Friday, December 4, 2015	
Blood & Respiration	O2/CO2 transport; WBCs & inflammation	Monday, December 7, 2015	
	Lung Ventilation: Compliance & Resistance	Tuesday, December 8, 2015	
	Lung Perfusion, Diffusion & Acid-Base	Thursday, December 10, 2015	
	Central & Peripheral Control of Ventilation	Friday, December 11, 2015	
	Hemorrhage	Monday, December 14, 2015	
	MODULE 3 EXAM 2	Tuesday, December 15, 2015	
	Special Topics in Systems Physiology	Wednesday, December 16, 2015	
	CHRISTMAS BREAK	2016	

CHAPTER ELEVEN **177**

TABLE 43 PPY511, 512, and 513 (2015–16), cont'd

Subtopic	Lectures	Dates	Course Title
Module 4: Endocrinology & Metabolism			PPY513 2 Credit Hours
Endocrinology	Endocrine Systems 1: Principles of Endo, Pituitary, and Adrenal	Monday, January 11, 2016	
	Endocrine Systems 2: Thyroid, Parathyroid, and Bone	Wednesday, January 13, 2016	
	Endocrine Systems 3: Growth and Metabolism	Friday, January 15, 2016	
	MARTIN LUTHER KING JR DAY	Monday, January 18, 2016	
Central Control	Principles of Energy Homeostasis	Wednesday, January 20, 2016	
	Hypothalamic/Autonomic Control of Appetite and Metabolism	Friday, January 22, 2016	
	Feedback System in Appetite Control	Monday, January 25, 2016	
	MODULE 4 EXAM 1	Wednesday, January 27, 2016	
GI Physiology	GI Hormones and Anatomy	Friday, January 29, 2016	
	Gastric Acid, Secretion, Pancreas and Bile	Monday, February 1, 2016	
	Transport, Absorption and Motility	Wednesday, February 3, 2016	
	Digestion and Energy Absorption I: From the Mouth to the Stomach	Friday, February 5, 2016	
	Digestion and Energy Absorption II: From Small Intestine and Out	Monday, February 8, 2016	
Energy Storage and Use	The Pancreas and Glucose Homeostasis	Wednesday, February 10, 2016	
	Liver Physiology and Cholesterol Synthesis/Transport/Degradation	Friday, February 12, 2016	
	Muscle and Protein Metabolism	Monday, February 15, 2016	
	Adipose Tissue and Lipid Metabolism	Wednesday, February 17, 2016	
	MODULE 4 EXAM 2	Friday, February 19, 2016	
	Special Topics in Endocrinology and Metabolism	Monday, February 22, 2016	
Module 5: Integrative Physiology & Pharmacology			PPY514 2 Credit Hours
	Hypertension: Etiology and Treatment	Wednesday, February 24, 2016	
	Heart Failure	Friday, February 26, 2016	
	Shock	Monday, February 29, 2016	
	Exercise and Altitude	Wednesday, March 2, 2016	
	Asthma	Friday, March 4, 2016	
	SPRING BREAK	March 7–12, 2016	
	Cancer, Chemotherapy, and Pain	Monday, March 14, 2016	
	Obesity and its Complications	Wednesday, March 16, 2016	
	Diabetes	Friday, March 18, 2016	
	Neurological and Neuropsychiatric Disorders	Monday, March 21, 2016	
	Special Topics/Student Presentations	Wednesday, March 23, 2016	
	GOOD FRIDAY	Friday, March 25, 2016	
	MODULE 5 AND/OR COMPREHENSIVE EXAM	Wednesday, March 30, 2016	
	EXPERIMENTAL BIOLOGY MEETING	April 2–6, 2016	
		May–July or June–August	

department decided to offer a grant-writing course as part of the formal requirements for the PhD. This course was largely designed and developed by Dr. Gina Yosten. Entitled Constructing Grant Applications in the Pharmacological Sciences (PPY514), the course carried one credit hour. Table 42´ describes the course as presented in Spring 2017.

Other aspects of the training program, such as qualifying and prelim exams, dissertations, and dissertation defense, remained essentially the same as described in detail in Chapter Ten.

Service Courses (2013–18)

The department continued to provide an undergraduate physiology course, Human Physiology (PPY254), Physiology in the Medical Anatomy and Physiology Program (MAPP), and Drugs We Use and Abuse (BLA245), taught by senior graduate students. Several other courses presented during the Westfall era were no longer offered, including Clinical Pharmacology for Physician's Assistants, General Physiology (PPYG), and Advanced Pharmacology in Primary Health Care Nursing (NRN508).

A new offering started in 2018 was a Human Physiology Summer Course (suPPY254). As was the regular PPY254 course, the summer course was taught entirely by Dr. John Chrivia and Dr. Joseph Baldassare and was available to senior and qualified high school students on their way to college. The course content and syllabus for both courses (PPY254 and suPPY254) were similar to what was presented previously in Chapter 10 and will not be repeated here.

Faculty Research Interests

Full-Time Tenure-Track Faculty

- **Michael Ariel, PhD, Washington University in St. Louis**
  Neuroanatomy and neurophysiology of the ocular system, including sensorimotor information processing from visual and vestibular inputs that control eye movements.
- **Thomas Burris, PhD, Florida State University**
  Use of chemical biology to characterize the physiological roles of nuclear receptors; development of drug targets of nuclear receptors for treatment of type 2 diabetes, heart disease, cancer, and Alzheimer's disease.
- **Andrew Butler, PhD, University of Auckland School of Medicine**
  Molecular biology and physiological role of melanocortin receptors (especially the melanocortin-3 receptor) and metabolic homeostasis; physiology and mechanism of the peptide hormone adropin.
- **Anutosh Chakraborty, PhD, Indian Institute of Chemical Biology**
  Impact of novel proteins in cell metabolism and survival; understanding the pathway of the inositol hexakisphosphate kinases (IP6Ks) in obesity, type 2 diabetes, hepatostenosis, and osteoporosis.
- **John Chrivia, PhD, University of Washington**
  Molecular biology and signal transcription factors, coactivator and nuclear proteins.
- **Bahaa El-Dien M. El-Gendy, PhD, University of Florida**
  Systems biology, drug development of anticancer agents, and automerism in molecular design of drugs.
- **Terrance M. Egan, PhD, Massachusetts Institute of Technology**
  Control of receptor- and voltage-gated ion channels using elecrophysiological and molecular biological approaches; biology of purinergic receptors, especially P2X subtype.
- **Colin A. Flaveny, PhD, Pennsylvania State University**
  Pharmacological targeting of LXR to treat prostate cancer.
- **Mark Knuepfer, PhD, University of Iowa College of Medicine**
  Autonomic pharmacology and physiology, central cardiovascular regulation, cardiovascular effects of cocaine and stress, mechanisms of hypertension.
- **Andrew Lechner, PhD, University of California, Riverside**
  Pulmonary physiology, acute lung injury, and the immunophysiology of sepsis.

- **Heather Macarthur, PhD, William Harvey Research Institute, Saint Bartholomew's Medical College**
  Vascular control mechanisms, endothelial mediators, sympathetic neurotransmission, pathophysiology of Parkinson's disease.
- **Wendi Neckameyer, PhD, Rockefeller University**
  Drosophila molecular neurobiology, function and regulation of dopamine and serotonin.
- **Daniela Salvemini, PhD, William Harvey Research Institute, Saint Bartholomew's Medical College**
  Novel approaches for therapeutic development, pharmacology of reactive oxygen/nitrogen species, role of sphingolipids and adrenosinergic signaling in pain mechanisms.
- **Willis (Rick) Samson, PhD, University of Texas Health Science Center**
  Neuroendocrine regulation of anterior pituitary hormone secretion, cardiovascular system regulation of fluid and electrolyte homeostasis, vasoactive peptides.
- **Mark Voigt, PhD, Saint Louis University**
  Biochemical and molecular neuropharmacology, molecular biology of serotonin and purinergic receptors, developmental neurobiology in zebrafish.
- **John K. Walker, PhD, Indiana University**
  Use of medicinal and synthetic organic chemistry, including structure-based and ligand-based drug design for development of new therapeutic agents and anti-inflammatory antibacterial and anti-cancer agents.
- **Gina Yosten, PhD, Saint Louis University School of Medicine**
  The roles of G protein-coupled receptors (GPCRs) in diabetes-associated microvascular dysfunction and GPCRs in the central circuits underlying obesity-associated hypertension; orphan GPCR function; GPR107 and GPR146 action and function.
- **Daniel S. Zahm, PhD, Milton S. Hershey Medical Center, Pennsylvania State University**
  Anatomy and function of the limbic system, the extended amygdala, and the nucleus accumbens; neurobiology of neurotensin and substance abuse.
- **Jinsong Zhang, PhD, University of Pennsylvania**
  Molecular biology of nuclear receptor corepressors, biology of histone deacetylases (HDACs), HDACs as drug targets in cancer, role of nuclear receptors in adipogenesis, inflammation, and lipid metabolism.

Research, Secondary, and Adjunct Faculty

- **Arindam Chatterjee**
  Design synthesis, purification, and characterization of different classes of organic compounds as nuclear receptor modulators.
- **Jane Cox**
  Neurotransmitter receptors, molecular biology, and developmental regulation of glutamate and P2X receptors.
- **Ian de Vera**
  Medicinal chemistry and synergistic regulation of coregulators and nuclear receptors.
- **Timothy Doyle**
  Role of sphingolipids and adenosinergic signaling in pain mechanisms.
- **John Edwards**
  Renal physiology, ion channel regulation of kidney transport.
- **Susan Farr**
  Behavioral pharmacology, mechanisms of neurodegeneration, aging and geropharmacology.
- **Kristine Griffett**
  Chronopharmacology and characterization of nuclear receptors.
- **Lamees Hegazy**
  Computer-based drug discovery, molecular modeling and simulations, structure–function relationships of biological macromolecules.
- **Scott Martin**
  Use of microchip-based analytical devices to study monoamines and peptide regulation.
- **James Willmore**
  Antiepileptic drug development and use, animal models of epilepsy.

TABLE 44 Extramural Support (2013–17)

Year	Extramural Support	NIH Rank in Pharmacology*
2013	$1,200,000	81
2014	$1,896,192	50
2015	$3,688,935	61
2016	$3,497,111	50
2017	$2,485,628	62

*Blue Ridge NIH Ranking

Extramural Support

Table 44 shows the total extramural support by full-time tenure-track faculty only in the Department of Pharmacology and Physiology for the period of 2013 to 2017. The numbers were provided from the Annual Chairs' Charter. Funding generated by research, secondary, or adjunct faculty was not included.

Faculty during the Burris and Voigt Eras

Near the time or soon after Dr. Westfall stepped down from the chair, several members of the existing department retired. Those included Mary Ellsworth, Al Stephenson, Michael Panneton, and Randy Sprague. Amy Harkins accepted the position of chair at the Department of Clinical Health Sciences in the Doisy School of Health Sciences, and Decha Enkvetchakul left the university.

Under the chairmanships of Dr. Burris and Dr. Voigt in the Department of Pharmacology and Physiology, there were nineteen tenure-track faculty members, eight research-track faculty, five faculty with secondary appointments, and five scientists with adjunct appointments.

Tenure-Track Faculty Who Transitioned from the Department under Dr. Westfall
(Year appointed is in parentheses.)
- Dr. Mickey Ariel (1993)
- Dr. Joseph Baldassare (1996)
- Dr. John Chrivia (1993)
- Dr. Terrance Egan (1992)
- Dr. Mark Knuepfer (1986)
- Dr. Andrew Lechner (1981)
- Dr. Heather Macarthur (1997)
- Dr. Wendi Neckameyer (1992)
- Dr. Daniela Salvemini (2009)
- Dr. Willis "Rick" Samson (1999)
- Dr. Mark Voigt (1994)
- Dr. Daniel "Scott" Zahm (1985)

Tenure-Track Faculty Recruited by Dr. Burris
- Dr. Andrew Butler (2014)
- Dr. Anutosh Chakraborty (2017)
- Dr. Bahaa El-Dien M. El-Gendy (2017)
- Dr. Colin A. Flaveny (2015)
- Dr. John Walker (2014)
- Dr. Gina L. C. Yosten (2015)
- Dr. Jinsong Zhang (2013)

Secondary, Research, or Adjunct Faculty Who Transitioned from the Department under Dr. Westfall
- Dr. Jane Cox
- Dr. Timothy Doyle
- Dr. L. James Willmore
- Dr. John Edwards
- Dr. Scott R. Martin
- Dr. Susan Farr
- Dr. William l. Neumann

Secondary, Research, or Adjunct Faculty Recruited by Dr. Burris
- Dr. Arindam Chatterjee
- Dr. Ian de Vera
- Dr. Sarbini Ghoshal
- Dr. Kristine Griffett
- Dr. Lamees Hagazy
- Dr. Marvin Meyers

Faculty during the Burris and Voigt Eras

(Years of service at Saint Louis University in parentheses.)

Jinsong Zhang, PhD (2013–present)

Dr. Zhang is a native of the People's Republic of China. He earned a BS in biochemistry from Nanjing University in 1988 and an MS in 1991 in biochemistry from the Shanghai Institute of Biochemistry and Cell Biology at the Chinese Academy of Sciences under the mentorship of Professor Wang L. Liu. This was followed by a PhD in biochemistry and molecular biology at the University of Pennsylvania under the mentorship of Dr. Mitchell A. Lazar in 1999. Dr. Zhang was a postdoctoral fellow in the laboratory of Dr. Robert G. Roeder in the Laboratory of Biochemistry and Molecular Biology at the Rockefeller Institute in New York City from 1999 to 2005. He was an assistant professor in the Department of Cancer Biology at the University of Cincinnati College of Medicine before joining the Department of Pharmacology and Physiology, Saint Louis University School of Medicine in 2013. Dr. Zhang's research is in the area of nuclear receptors and cancer biology, specifically the mechanism of transcriptional regulation by nuclear receptors and G protein transcriptional factors. He is well supported by grants (both R01 and R23) from the NIH. He has published over thirty peer-reviewed papers and has given numerous talks at national and international meetings and seminars at other institutions. He is active as a grant and journal reviewer and has served on numerous PhD qualifying and thesis committees. Dr. Zhang has mentored two PhD students and five postdoctoral fellows to date, as well as numerous rotation and undergraduate students. He has also been active on departmental and university committees.

Andrew A. Butler, PhD (2014–present)
Dr. Butler is a native of Cambridge, New Zealand. He earned a BS in zoology (1st Class Honors) from the University of Canterbury in 1991. This was followed by a PhD in pediatric endocrinology at the University of Auckland School of Medicine in 1995. He was a Fogarty International Fellow at the National Institute of Diabetes and Digestive and Kidney Diseases (NIDDK) in Bethesda, Maryland, from 1995 to 1997, after which he took a postdoctoral fellowship at the Vollum Institute of Oregon Health & Science University from 1998 to 2001. He was appointed assistant professor (tenure track) at the Pennington Biomedical Research Center at Louisiana State University from 2001 to 2007. He became director of the Animal Metabolism and Behavior Core and associate professor (with tenure) at the Clinical Nutrition Research Unit of the Pennington Biomedical Research Center from 2007 to 2009. From 2009 to 2014, Dr. Butler was an associate professor (with tenure) at the Scripps Research Institute. Dr. Butler was appointed professor (with tenure) in the Department of Pharmacology and Physiology at the Saint Louis University School of Medicine in 2014.

Dr. Butler's research involves neuroendocrinology and the role of novel neuropeptides in aging, cognitive disorders, ischemia, stroke, and circadian rhythms. His research has been supported by R01s from NIDDK and NINDS, private foundations such as the American Diabetes Association and the Lottie Caroline Hardy Charitable Trust, and pharmaceutical companies such as Sanofi-Aventis. As of late November 2017, Dr. Butler had published fifty-one peer-reviewed papers, five book chapters, twenty-six invited monographs, and seventy-nine invited presentations or abstracts. To date, he has mentored ten postdoctoral fellows and two undergraduate students and has served on four PhD dissertation committees. He has received several honors and awards, including the Lilly Scientific Achievement Award from the Obesity Society in 2008 and the Junior Faculty Award from the American Diabetes Association.

John K. Walker, PhD (2014–present)
Dr. Walker is a native of the metropolitan St. Louis area. He received a BS in chemistry from Southern Illinois University at Edwardsville in 1990 followed by an MS in organic chemistry from that institution in 1992. He earned a PhD in organic chemistry from Indiana University in 1999 working under the mentorship of Dr. Paul A. Grieco. He was a research scientist at Monsanto's, Searle Pharmaceutical Division from 1998 to 2001, and followed this position with several others in private industry, including senior scientist at Pharmacia Global Research from 2001 to 2003, senior principal scientist at Pfizer Pharmaceutical Global Research from 2003 to 2007, and associate research fellow there from 2007 to 2010. From 2010 to 2014, Dr. Walker was a research assistant professor and director of medicinal chemistry at the University of Missouri-St. Louis. He was appointed to his present position of assistant professor in the Department of Pharmacology and Physiology at the Saint Louis University School of Medicine in 2014, and was promoted to Associate Professor with Tenure in 2019. He has a secondary appointment as

assistant professor in the Department of Chemistry at Saint Louis University. Dr. Walker is a member of several professional societies, including the American Chemical Society's Division of Medicinal Chemistry and the National Academy of Inventors. He has been awarded ten patents with twelve pending, written twenty-seven peer-reviewed papers, attended nine invited research seminars or conference presentations, and written seven abstracts. His research has been generously funded by grants from the NIH and the Department of Defense. To date he has mentored four postdoctoral fellows and three undergraduate students. He actively participates in the teaching and training of medical, graduate, and undergraduate students and has served on numerous university, medical school, and departmental committees. He is co-principal program director of a T32 training grant in Pharmacological Sciences.

Anutosh Chakraborty, PhD (2017–present)
Dr. Chakraborty is a native of India. He received a BS in zoology, chemistry, and botany (First Class) from Burdwan University in India in 1995 and an MS in zoology (First Class) from the same school in 1997. This was followed by a PhD from the Indian Institute of Chemical Biology in 2005. Dr. Chakraborty was a postdoctoral fellow at the Johns Hopkins Medical School from 2005 to 2008 and a research associate and instructor at that institution from 2008 to 2012. He joined the Scripps Research Institute in 2012, first as an assistant professor from 2012 to 2017 and then as an associate professor in 2017. Dr. Chakraborty was appointed associate professor in the Department of Pharmacology and Physiology at the Saint Louis University School of Medicine in August 2017. Dr. Chakraborty's research deciphering the impact of novel biomolecules in cell signaling, metabolism, and survival has implications in various diseases including obesity, diabetes, cardiovascular disease, cancer, and psychiatric and neurodegenerative disease. To date he has published twenty peer-reviewed papers and twenty-three abstracts. He has given seventeen invited lectures and has one patent. His research has been supported by the NIDDK and he has served on scientific review committees and as a reviewer for several international journals. He has mentored two postdoctoral fellows, two master's degree students, summer interns, and high school students. Dr. Chakraborty also participates in teaching activities at the various institutions he has served.

Bahaa El-Dien M. El-Gendy, PhD (2018)
Dr. El-Gendy is a native of Egypt. He earned a BS in Chemistry from Benha University in Egypt in 1999, followed by an MS in organic chemistry from the same university in 2003. He earned a PhD in chemistry from the University of Florida in Gainesville in 2010. He was a postdoctoral associate in the Department of Chemistry at the University of Florida from 2010 to 2012, after which he held the position of research associate at the Translational Research Institute of the Medicinal Chemistry Department at Scripps Research Institute from 2012 to 2013. He held the position of assistant professor of organic and medicinal chemistry in the Department of Chemistry at Benha University in Egypt from 2011 to 2016 and associate professor in that department from September 2016 to the present. He was appointed assistant professor of medicinal chemistry in the Department of Pharmacology and Physiology at the Saint Louis University School of Medicine in June 2017. His research is in systems biology, medical chemistry, chemical biology, drug development of anti-cancer drugs, and tautomerism in molecular design. To date he has published twenty-five peer-reviewed articles and has given numerous invited seminars and workshops. He has earned numerous honors and awards, including the Presidential Medal of Excellence (first class) for outstanding contributions from the president of Egypt in 2017. Dr. El-Gendy left the department in 2018 to join Dr. Burris at the St. Louis College of Pharmacy.

Colin Ashton Flaveny, PhD (2014–present)
Dr. Flaveny is a native of the Bahamas. He earned a BSc in biology from Howard University in 2003. He followed this with a PhD in molecular toxicology from Pennsylvania State University in 2010 under the mentorship of Dr. Gary H. Perdew. He was a postdoctoral research associate at the University of Miami Miller School of Medicine from 2010 to 2012, and then held a similar position at the Scripps Research Institute in 2012 and 2013. He conducted postdoctoral research

in the Department of Pharmacology and Physiology at the Saint Louis University School of Medicine from 2013 to 2014 under the mentorship of Dr. Thomas Burris. He was appointed assistant research professor in the department in 2014, and in 2015, he was promoted to assistant professor (tenure track) in the department. Dr. Flaveny's research is in cancer biology and drug discovery, specifically, the pharmacological targeting of LXR to treat prostate cancer as well as REV-ERB ligands for the treatment of anxiety disorders. He has received funding as principal investigator from the NIH, the Department of Defense, and private institutions. To date he has published more than twenty-two peer-reviewed papers, three review articles, and two book chapters. He has also given approximately twelve conference and meeting presentations. He is actively mentoring five PhD students and is participating in medical and graduate school courses. He has received several outstanding research awards from the Society of Toxicology.

Gina L. C. Yosten, PhD (2015–present)
Dr. Yosten is a native of Oklahoma. She earned her BS in zoology-biomedical sciences (summa cum laude with distinction) from the University of Oklahoma in 2004. She followed this accomplishment with a PhD in pharmacological and physiological sciences from the Saint Louis University School of Medicine in 2010 under the mentorship of Dr. Willis (Rick) Samson. She was a postdoctoral fellow in the Department of Pharmacological and Physiological Science from 2010 to 2013 and was appointed research assistant professor in 2013 and assistant professor (tenure track) in 2015 in the Department of Pharmacology and Physiology. She has received many awards in her young career including membership in the Phi Beta Kappa Honor Society, five awards from the American Physiological Society, two from the Endocrine Society, and numerous awards for research presentations nationally and internationally. She has given ten invited presentations at national and international meetings and symposia and to date has published over twenty-five papers in peer-reviewed journals, four invited review articles, and three book chapters, as well as numerous abstracts. She has served on four society committees: the Science Policy, Joint Program, and Endocrinology and Metabolism Section Steering Committees of the American Physiological Society (APS); and the Training Subcommittee for the Federation of American Societies for Experimental Biology (FASEB). She also serves on two editorial boards. She has received funding from the NIH and private foundations. She teaches in medical school modules, core graduate student courses, and pharmacology and physiology graduate courses. She designed a grant-writing course for pharmacology and physiology graduate students and serves as its course director. She has also mentored several PhD students. She has been chosen to serve as the Editor-in-Chief of the *American Journal of Physiology: Regulatory, Integrative and Comparative Physiology* in June 2020.

Mark Knuepfer, PhD
See page 80.

Andrew John Lechner, PhD
See page 60.

Terrance M. Egan, PhD
See page 146.

John Chrivia, PhD
See page 147.

Mark Massie Voigt, PhD
See page 148.

Joseph Baldassare, PhD
See page 148.

Heather Macarthur, PhD
See page 149.

Wendi Neckameyer, PhD
See page 147.

Willis Kendrick "Rick" Samson, PhD
See page 150.

Daniela Salvemini, PhD
See page 153.

Daniel Scott Zahm, PhD
See page 153.

Michael Ariel, PhD
See page 154.

Stacy Smith (2013–present)

Special recognition is given to Stacy Smith, who has served as business manager of the department from 2013 to the present. Stacy was born and raised on the east side of the St. Louis metropolitan area. After spending time in Florida, Stacy returned to St. Louis and was appointed business manager in 2013 by Tom Burris. Stacy served with distinction during the tumultuous years from the end of the Westfall era through the Burris, Voigt, and Salvemini years and has helped the transitions take place more smoothly.

PhD Graduates and Students during the Burris and Voigt Eras

The graduate program that was developed during the Westfall era has been maintained under the leadership of Dr. Burris, Dr. Voigt, and Dr. Salvemini. The numbers of students entering the program and graduating have also been similar. Students receiving their PhD while Dr. Burris and Dr. Voigt were chair are listed below. This list overlaps somewhat with students listed in Chapter Ten, since they entered the program while Dr. Westfall was chair.

PhD Graduates

- *2014 Mollisa Elrick
- *2014 Amanda (Finley) Ford
- *2014 Chloe (Brant) Tollock
- *2014 Heather Lavezzi
- *2014 Jennifer Richards
- *2014 Holly Pope
- *2015 Julienne Touchette
- *2015 Robert Adams
- *2016 Lauren Stein
- *2016 Carrie (Malroy) Wahlmann
- *2017 Rhett Reihard
- *2017 Katie Stockstill
- **2017 Ryan Welch
- **2018 Parag Bhatt
- **2018 Laura Banks
- **2018 Deborah Roby
- **2019 Christopher Haddick
- **2019 Jacqueline Rossiter
- **2019 Monideppa Sengupta
- **2020 Katherine Beaufort-Carpenter

*Students entering the program while Dr. Westfall was chair.

**Ryan Welch was the first student to graduate following entry into the program after Dr. Burris became chair.

Current PhD Students

Name	Mentor
Abuirqeba, Suomia	Flaveny
Breden McInerny, Kathryn	Salvemini
Harada, Caron	Salvemini
Haubner, Jacob	Chakrahorty
Kummer, Katherine	Macarthur
Mahmood, Ammer	Yosten
Majidi, Shabnam	Flaveny
Michalski, Stephanie	Egan
Msengi, Naomi	Chakraborty
Mufti, Fatima	Salvemini
Pavlack, Monica	Nguyen
Schafer, Rachel	Salvemini
Schnell, Abigayle	Samson/Yosten
Schoepke, Emmalie	Walker
Wong, Chu Fung (Johnny)	Chakraborty

MD/PhD Students

Name	Mentor
Grote, Stephen	(GY) Yosten
Murray, Meghan	Flaveny
Steinauer, Nickolas	Egan/Zhang
Goedland, Monica	Farr

Transition from the Burris Era to the Voigt Era and the Future of the Department

Prior to the departure of Dr. Thomas Burris as chair of the Department of Pharmacology and Physiology there was a great deal of anxiety and unrest in the department about its future. Dean Behrns had proposed a plan that would eliminate individual departments and replace them with twelve institutes. This would have resulted in departmental faculty choosing or being placed in separate institutes, meaning some colleagues would be separated. It would have also meant the disappearance of the Department of Pharmacology and Physiology. It was also unclear what effect this would have on graduate programs. Dr. Burris thought that this would be very disruptive of his plans for chemical biology and

FIG. 43 Dr. Mark Voigt

drug discovery in that there wasn't an institute that had these topics as the major objective. He decided that he would be better off at another institution and accepted a position as assistant to the president of the St. Louis College of Pharmacy, with joint appointment in two departments at Washington University. He resigned from Saint Louis University in February 2018 and moved his laboratory to the St. Louis College of Pharmacy soon after. Dr. Mark Voigt was appointed interim chair of the department for the second time in February 2018. In the meantime, the plan to move from a department model to an institute model was abandoned. The position of vice dean was established, and the duties of the dean's office were split between Dean Behrns and the new vice dean, Dr. Bob Wilmott. Dr. Wilmott was the chair of the Department of Pediatrics. Dr. Willmott was appointed permanent dean in the fall of 2019.

TABLE 45 Department of Pharmacology and Physiology Faculty Meeting Agenda, August 2, 2018

I. POINTS OF INFORMATION
- people in the press:
  - Gina in SLBJ
  - Daniela in SLM
- Gerry Wilken will be contact person re: equipment

II. OLD BUSINESS
- course updates
- committee update (see attached list):
  - which ones do we need / refreshing membership on some
- chalk talk scheduling
- faculty evaluation policy (see attached notes)

III. NEW BUSINESS
- new Chemical Biology program (attached)

Faculty Who Transitioned from the Department under Dr. Burris

Except for the members of Dr. Burris's laboratory and several who had positions as research faculty, all of whom moved with Dr. Burris to the St. Louis College of Pharmacy, all other members of the department remained at Saint Louis University. These included Dr. Ariel, Dr. Baldassare, Dr. Butler, Dr. Chakraborty, Dr. Chatterjee, Dr. Chrivia, Dr. Cox, Dr. de Vera, Dr. Doyle, Dr. El-Gendy, Dr. Egan, Dr. Flaveny, Dr. Hegazy, Dr. Knuepfer, Dr. Lechner, Dr. Macarthur, Dr. Salvemini, Dr. Samson, Dr. Walker, Dr. Yosten, Dr. Zahm, and Dr. Zhang. The activities of research, teaching, training, and service carried out by the department were continued under the excellent leadership of Dr. Mark Voigt as the interim chair until June 2019, when Dr. Salvemini was appointed interim chair, and the future of the department once again looks very promising.

TABLE 46 Department of Pharmacology and Physiology Retreat Agenda, May 9, 2019

Departmental Business (30 minutes)
1. FY20 Budget
2. Statistical Overview

Research (30 minutes)
1. Metabolic Core
2. Major Equipment
3. Grant Submissions

Teaching (45 minutes)
1. Graduate Student Education
   - Handbook
   - Student involvement
   - T32
   - Core Program
2. Medical Student Education
   - Curriculum changes
3. Undergraduate Education
   - Chemical Biology program
   - PPY254

School/University Issues (45 minutes)
1. Faculty Compensation Plan
2. Departmental Website
3. Budget Issues
4. Postdoctoral Salaries
5. Staff
6. Pharm Phys moving forward

Faculty Meetings and Retreats

Dr. Voigt continued having faculty meetings on a monthly basis, as well as yearly retreats. Table 45 shows a representative agenda of a faculty meeting, and Table 46 shows the 2019 retreat agenda.

Update of Departmental Committees

Dr. Voigt also decided to decrease the number of committees that existed under Dr. Burris. Below is the most recent listing of departmental committees.

Pharmacology and Physiology Committees (as of June 18, 2019)

Seminar Committee
- Butler, Chair
- Yosten
- Walker
- Zhang

Graduate Education Steering Committee
- Egan, Chair
- Lechner
- Yosten
- Zhang
- Macarthur
- Salvemini
- Flaveny

Tenure/Promotion Committee
- Samson, Chair
- Salvemini
- Egan
- Lechner

Communication (Website, Facebook, etc.) Committee
- Zhang, Chair
- Butler

Preliminary Exam Standing Committee
- Macarthur, Chair
- Lechner
- Zhang
- Walker
- Ariel

Departmental Executive Committee (Policies) Committee
- Voigt, Chair
- Lechner
- Salvemini
- Samson
- Stacy Smith

Journal Club Committee
- Zahm, Chair
- Egan
- Samson
- Flaveny

Proposal for Faculty Evaluation

This document describes a system of evaluation for the faculty of the Department of Pharmacoloy and Physiology as proposed by Dr. Voigt. The purpose of this evaluation is to provide a system for the appraisal, development and documentation of each faculty member, with the purpose of promoting the intellectual growth of individual faculty while enabling the department to fulfill its collective goal. It will be open and transparent to all being evaluated, so that faculty members will clearly know what is expected of them from each of the four categories of evaluation, as formalized and agreed to in their meetings with the Department Chair prior to the start of each academic year. Comparing each faculty member's efforts will enable the Department Chair to determine whether that faculty member has fulfilled or exceeded the minimum effort expected for that academic year, based on the agreements made prior to that year.

This evaluation requires the active participation of each faculty member, in concert with the Department Chair, to create his/her annual professional expectations and then to provide evidence a year later of his/her achievements based on those professional activities. Faculty activities under review are divided into the categories of formal teaching, mentoring, research, and service/administration.

This process was developed by a subcommittee of the department, and will be submitted for approval to the faculty and Department Chair.

ANNUAL ASSESSMENT PROCESS OF DEPARTMENTAL FACULTY IN PHARMACOLOGY AND PHYSIOLOGY

Each faculty member will meet in the spring with the Department Chair to discuss his/her goals for the upcoming academic year, attempting to quantify those goals in terms of the metrics stated to the right. The result will be an annual 'faculty goals' document signed by both (a copy of which will be retained in the faculty member's file). An important component of that document is the statement of Department resources that the Chair agrees to provide to the faculty member in support of the proposed goals.

The document will specify what range of success meets a faculty member's minimal performance as well as provide additional levels of performance to serve as incentives to achieve a superior performance. In this way, small cyclical or periodic fluctuations in activities that are inevitable in academic life will not prove detrimental to a faculty member. In the autumn of the following year, each faculty member will compile an annual report summarizing his/her accomplishments for the preceding academic year, specifying the level of success in each relevant specific category stated in the faculty goals document. This document will be submitted to the Department Chair and an evaluation meeting will then be scheduled. At the evaluation meeting, the faculty member and the Department Chair will discuss how the 'goals' were attained or exceeded and any adjustments to be made by the Department Chair. If there are serious discrepancies between the chair's and the faculty member's analyses, then it is advised that additional meetings be held by the Department Chair and the faculty member in an attempt to resolve such differences. Written adjustments will be finalized by the Department Chair and provided to the faculty member. This document will also be used by the Department Chair for the planning of the upcoming fiscal information required by the Dean for future budgetary planning.

Potential totals for evaluative categories:
Meets Expectations: 100-125 pts
Exceeds Expectations: 126-150 pts
Outstanding: >150 pts

PROPOSED METRICS OF RESEARCH EVALUATION
Acceptable point total = 50
- *Publications:*
  - Peer reviewed papers, number, impact factor (5 - 15 points per paper)
  - Invited and peer reviewed book chapters and review articles (10 - 15 points for each)
- *Collaborative efforts:*
  - Internal and external (10 points)
- *Presentations:*
  - Regional (2 points)
  - National (10 points)
  - International (15 points)
- *Invited seminars:*
  - Internal (SLU)/regional 5 points
  - National, 10 points
  - International, 15 points
- *Grant submissions:*
  - Industry (5 points), Foundations (5 points), NSF (10 points), NIH (10 points R21 or SBIR/STTR, 20 points RO1)
- *Grant awards:*
  - Duration, indirect cost recovery (25 - 50 points)
- *Other awards:*
  - To be discussed during evaluation since it will depend on the type of the award

Expectations of typical faculty member:
2 papers   10 pts
1 coll. effort   10 pts
1 poster @ nat'l   10 pts
1 RO 1 submission 20 pts
                   50 pts

PROPOSED METRICS OF TEACHING
Acceptable point total = 25
NOTE: there will be a multiplier of 0.5/1/1.5x used based on student evaluations
- *Undergraduate students*

- Course Director: 5-40 pts. Points awarded based on the extent of administrative activities.
- Lecture or small group interaction: 1 point per hour. Development of new lecture: 2 points per hour.
- Non-course director administration activities (1-5 points). Points awarded based on extent of administrative activities.
- *Graduate students*
  - Course Director: 5-40 pts Points awarded based on extent of administrative activities.
  - Lecture or small group interaction: 1.25 points per hour. Development of new lecture: 2.25 points per hour
  - Non-course director administration activities (1-5 points). Points awarded based on extent of administrative activities.
- *Medical students*
  - Course Director: 5-40 pts. Points awarded based on the extent of administrative activities.
  - Lecture or small group interaction: 1.5 points per hour. Development of new lecture: 2.5 points per hour
  - Non-course director administration activities (1-5 points). Points awarded based on the extent of administrative activities.
- *Administrative activities include:*
  - Syllabus preparation
  - Exam preparations and proctoring. Office hours and student appointments
  - Management of SLU Global website or other class related websites
  - Homework assignments, including preparation time and time spent answering student questions. Development/editing written course materials
  - Letters of recommendation
  - New course development

Expectations of typical faculty member:

4 hrs med. students	6 pts
12 hrs grad students	15 pts
	21 pts

PROPOSED METRICS OF SERVICE/ADMINISTRATION EVALUATION

Acceptable point total = 15

- University-wide standing committee chair (15 pts)
- Medical School standing committee chair, Center director, Program director, Task Force director (10 pts)
- Departmental standing committee chair or graduate program director (10 pts)
- Chair of community service organization related to one's professional expertise (5 pts)
- National professional organization's leadership role or editorial board (15 pts)
- Member of national organization or federal/state/foundation grant review panels (15 pts)
- Membership in select time-intensive committees (e.g. IACUC, 10 pts)
- Member of community service organization related to one's professional expertise (2 pts)
- Organize Journal Club or Lecture Series at departmental, school or University level (5 pts)
- Ad-hoc reviewer for journal articles or grant proposals (each 2 pts)
- Member of departmental, school or University level committee (each 5 pts)
- Liaison for the department at school or University level committees (each 2 pts)
- Member of Faculty Senate (5 pts)
- President/member exec. comm of Faculty Senate (15 pts)

(points prorated for short-term or ad-hoc committees)

Expectations of typical faculty member:

1 dept'l comm	5 pts
1 med sch comm	5 pts
2 ad hoc reviews	4 pts
	14 pts

PROPOSED METRICS OF MENTORSHIP
Acceptable point total = 10
- *Internal (direct mentorship in your lab):*
  - Graduate student (10 points)
  - Postdoctoral (15 points)
  - Undergraduate research assistants (5 points per student)
  - High school students (5 points per student)
  - Differentials: inclusion on abstracts and papers, awards won, thesis completion, meeting presentations
- *External (mentorship of others):*
  - Preliminary/thesis committees (2/5 points per committee)
  - Collaborator/consultant (students, post docs, junior faculty) (2 points)
  - Differentials: departmental, SLU, national, international, extent of effort
- *Hosting:*
  - Rotations (5 points per student)
  - Visiting faculty/sabbaticals (5 points)
  - Seminars (2 points per seminar)
- *Jesuit Mission:*
  - Inclusion/encouragement/mentorship of students in outreach (e.g., Science Fair, STARS, Speaker for Science) (5 points for each)
  - Faculty participation in above (5 points for each)

Expectations of typical faculty member:

1 prelim comm	2 pts
2 seminars	4 pts
1 collaboration	2 pts
1 undergrad/HS student	5 pts
	13 pts

Teaching Activities

The teaching activities of the department to medical and graduate students and in service courses continue as described previously under the Burris era. A summary of the overall teaching activities (hours) in 2018 and 2019 are summarized in Table 47. An additional program beginning in Fall 2019 is an undergraduate BS program in chemical biology and pharmacology (CBP). This new degree program is a joint effort of the Department of Pharmacology and Physiology and the Department of Chemistry and will involve developing courses in molecular pharmacology and systems pharmacology. These courses will be similar to PP511, 512, and 513.

Table 48 includes a summary of the departmental service and reputational efforts for FY 2019, and Table 49 gives a research overview for 2018–19. Finally, Table 50 depicts the strategic plan developed by Dr. Voigt and presented at the departmental retreat in May 2019.

Epilogue and the Future of the Department of Pharmacology and Physiology

While the substance of this book was brought to a close at the end of 2018 and beginning of 2019, several events occurred in 2019 that have resulted in an epilogue. Under the able leadership of Dr. Mark Voigt, the teaching and training activities of the department were flourishing by May of 2019, as can be seen in Tables 47. Likewise, the research productivity of the department was doing very well. There were numerous extramurally funded and exciting research projects under way, as seen in Table 49, and the department seemed quite stable financially. However, by June 2019, several things occurred that suggested the future of the department, as such, was uncertain.

Although Dr. Voigt was doing an outstanding job in stabilizing and moving the department forward, the departure of Dr. Burris left the department in a vulnerable position. This was no doubt highlighted by the fact that the university and the School of Medicine had made a considerable investment in Dr. Burris and the department in the form of a substantial set-up package. Since the departure of Dr. Burris in February

TABLE 47 Pharmacology and Physiology Teaching Overview (2018–19)

Medical School Teaching

MS 1	147 hrs	
MS 2	104 hrs	} 251 hrs

Four Course Directors
- Pharmacology Module
- Cardiovascular Module
- Respiratory Module
- Endocrine Module

Graduate Student Teaching

PPY 511	18 hrs	
PPY 512	50 hrs	
PPY 513	43 hrs	} 156 hrs
PPY 514	45 hrs	
BBSC 501	3 hrs	
BBSC 502	18 hrs	
BBSC 503	55 hrs	} 160 hrs
BBSC 504	18 hrs	
BBSC 592	66 hrs	
ANAT 5400	50 hrs	} 50 hrs

Undergraduate Teaching

BLA 245	50 hrs	
PPY 254	42 hrs Lecture	} 447 hrs
	355 hrs Tutorial	

TABLE 48 Departmental Service and Reputational Efforts (2018–19)

Memberships
- School of Med Committees/Position — 26 (Including Biomed Core Director)
- University Committees — 14 (Including IACUC Chair, Center for Neuroscience)
- National Study Sections — 9
- Editorial Boards — 14 (Including Editor in Chief, AJP Journal)
- National Leadership Positions — 5 (Including Council, APS)

Invitations
- Reviews/Book Chapters — 9
- National Seminars — 6
- National Symposia — 7
- International Symposia — 7

Patents — 4

Collaborations — 20
(unique institutions)
- Include 11 in USA (MD Anderson, Child Hosp. Cincinnati, NIH, UCLA: Univ Arizona, Univ Kentucky, Univ Oklahoma, Univ Penn, SUNY Stony Brook, YCU, WUSL StCOP)
- Outside USA (Australia; Canada; Chile; England; France; Germany, Italy)

TABLE 49 Departmental Research Overview (2018–19)

Current External Funding
- NIH R01s — 9
- NIH R21s — 2
- NIH subcontracts — 2
- T32 NIH training grant — 1
- DOD Grant — 1
- Industry Grant — 1
- Foundation Grants — 2
- Total — 18

Pending Grants
- NIH — 14
- Foundation — 2
- Siteman/ICTS — 5

External Applications
- YTD — 30
- NIH — 15
- Foundation — 7
- DOD — 3
- Siteman/ICTS — 5

Internal Applications
- YTD — 14
- RGF — 9
- PRF — 5

TABLE 50 Pharmacology and Physiology Strategic Plan (as of 2019)

Increase external funding
target is 3 new R01s and 1 R21 in FY20

Bring Mouse Metabolic Phenotyping Core online
this should happen in May '19

T32 training grant submission
due date September 25, 2019

Individual faculty development plans
focus on early and mid-career first

Manage staffing changes
ongoing

Space reallocation re: Clinical Pathology on 4th flr Schwitalla
ongoing

Build departmental reserves
ongoing

TABLE 51 Total Number of Ph.D. Graduates Mentored by Faculty in Pharmacology and Physiology at Saint Louis University*

Degree Topic	Number	
Pharmacology/Physiology	91	
Physiology	50	177
Pharmacology	36	
Cell and Molecular Biology	11	
Biology	2	16
Pathology	2	
Biochemistry and Molecular Biology	1	
Total	193	

*through 2019

2018, there had been little talk by the administration about a permanent chair for the Department of Pharmacology and Physiology. Dr. Voigt's title during the entire time he served was interim chair. Rumor had circulated for a considerable time that some members of the administration wanted to consolidate the basic science departments into a single unit. The fact that so many permanent chairs in clinical departments were vacant suggested that a permanent chair in the Department of Pharmacology and Physiology might have to be put on hold.

It was well known that Dr. Voigt wanted to retire in 2020, and at the monthly meeting between Dean Wilmott and Dr. Voigt in May 2019, the dean suggested that this might be accelerated. The dean asked Dr. Voigt if he would be willing to retire earlier and suggested that Mark take the retirement package known as the VERP (Very Early Retirement Package). Dr. Voigt assumed that the school "no longer wanted him to be interim chair and to be gone." He agreed to the buyout package and quickly retired on June 31, 2019.

Rumors were again rampant that the administration of the university and the School of Medicine would dissolve the Department of Pharmacology and Physiology, merge it with the Department of Biochemistry and Molecular Biology, and eventually consolidate all the basic sciences into one unit or department. The entire department quickly requested a meeting with Dean Wilmott to discuss its future and to emphasize the importance of remaining a separate department. Dr. Salvemini, Dr. Walker, Dr. Egan, and Dr. Yosten compiled a list of reasons for maintaining an independent Department of Pharmacology and Physiology. This list is summarized below.

1. Identity. The Department of Pharmacology and Physiology offers the faculty a sense of identity within the disciplines that would not exist in a merged department. This identity is a crucial part of our reputation with funding agencies, such as the National Institutes of Health. Thus, maintaining the department will enable our faculty to continue to obtain extramural funding based on our expertise in the pharmacological and physiological sciences. Furthermore, our T32 training grant in Pharmacology is dependent upon the existence of a Department of Pharmacology and Physiology. Dissolution of the department would understandably lead to loss of our T32 funding.

2. Maintenance of a Diversified Research Portfolio. The Department of Pharmacology and Physiology is made up of faculty with diverse expertise that is distinct from that of other departments. The continued existence of the department ensures that this diversity is maintained. Merging the department with other departments would lead to a reduction in research foci and a resultant reduction of applicable funding opportunities for our scientists. It is essential to maintain a diversified research portfolio in order to maximize the amount of extramural funds obtainable by SLU faculty. It should be noted that extramurally funded biomedical research conducted in PPS is by nature translational, thus fitting well the mission of the NIH. A department merge would dilute and possibly divert that translational focus.

3. Education. The Department of Pharmacology and Physiology has a rich tradition of providing quality education to medical, graduate, and undergraduate students. Our faculty are course directors for multiple medical school modules, which contributes heavily to the tuition dollars obtained by the School of Medicine. Our success in education stems from coordinated efforts between our faculty that would be disrupted if the department were dissolved. Thus, our ability to provide quality education for the School of Medicine would be greatly inhibited. In addition, our department has a history of producing high-quality scientists that are an excellent advertisement for SLU. Like the faculty, our students base their scientific identity on the department, and we should not strip them of that identity.

4. Culture. The culture of our department is important for fostering collaborations and professional development between and for our faculty and students. Because of our established culture, members of our faculty have obtained (or will soon obtain) collaborative extramural grants. Likewise, we have an established system for mentorship of junior faculty. Moreover, our department culture is essential for maintaining the well-being of our faculty, staff, and students. Dissolution of the department would lead to employee dissatisfaction and potentially loss of funded faculty.

5. Resources and Facilities. Our department houses equipment and other resources that are essential and unique to the researchers in Pharmacology and Physiology. These resources must be maintained in order to support the research mission of the department. Assimilation of these resources into a larger entity, such as a merged department, would likely lead to disruption of resource management. Likewise, many Pharmacology and Physiology faculty require specialized research space for animals and chemical synthesis hoods, for example. These specialized lab spaces would be difficult and expensive to replicate in another space (i.e., the DRC) and are necessary for the continued productivity of Pharmacology and Physiology faculty. Thus, a department merge and/or move would significantly curtail research productivity and cost the university significant funds.

In addition to these reasons for maintaining a Department of Pharmacology and Physiology, the graduate students of the department also prepared a paper presented to the the dean with arguments and reasons for maintaining a separate department and training program. This is summarized as follows and was presented by Katie Carpenter on behalf of the students at the meeting with the dean.

Department of Pharmacology and Physiology Statistics

1. Student Body. The Department of Pharmacology and Physiology has the largest student body in the School of Medicine. Currently we have four MD/PhD students and fifteen PhD students (fourteen entered from the Core Program and one direct entry).

2. Pharmacology T32 Training Grant. The Department of Pharmacology and Physiology is the only SLU department to hold a NIH T32 training grant. This grant has been awarded and renewed in the department for over twenty-eight consecutive years. The training grant currently supports four graduate students per year. It also provided a student-led diverse careers seminar series. Current graduate students reach out to professionals with PhDs who are currently in a career that is not a traditional professorship. These professionals come to the department and meet with students to discuss their career paths.

3. BLA245 Drugs We Use and Abuse. All third-year graduate students in the Department of Pharmacology and Physiology teach at least one lecture in the undergraduate course BLA245. This course is offered to non-science undergraduate majors (about fifty students per year) and covers everything from basic physiology and pharmacology to disease states such as cancer, Alzheimer's disease, heart disease, etc. It also discusses commonly abused drugs such as opioids, hallucinogens, and alcohol. No other department in the School of Medicine consistently offers graduate students the opportunity to obtain experience in formal teaching.

4. Student Success. The Department of Pharmacology and Physiology has a long track record of student success. Approximately 99 percent of students that enter the department complete their PhD. Over the years, the department has graduated 177 PhDs in Pharmacology/Physiology, 126 of these since 1986. Sixteen others were mentored by faculty in the Department of Pharmacology and Physiology, bringing the total to 143 individuals.

After leaving the department, alumni are continually successful. To date our students are continuing to be successful as postdocs at prestigious institutions (Harvard University, Stanford University, Vanderbilt University, Salk Institute, etc.). Many have professorships at universities (University of Virginia, University of Florida, University of Alabama at Birmingham, Washington University, etc.) or are research scientists (Endocyte-Novartis, Merck, etc.), senior medical liaisons, and directors of medical writing, among other careers.

5. Travel Provision and Thomas C. Westfall Fellowship. The Department of Pharmacology and Physiology is the only School of Medicine department to guarantee travel to one conference per year per student. This allows graduate students to present their research at prestigious conferences such as Experimental Biology, Society for Neuroscience (SFN), Keystone Symposia, Cell Symposia, the American Association of Immunologists' annual conference, and others. In addition to travel provisions, the department offers the Thomas C. Westfall Fellowship. This fellowship is awarded annually to one graduate student based on a short podium talk discussing their research. This award has allowed students to attend both national and international meetings.

6. Current Student Awards and Success. Current graduates have received awards from SLU (GSA, Brennan Summer Research Award, GSA Paper Presentations Award) and nationally (Department of Defense Predoc Horizon Grant, American Physiology, Society Research Recognition Award, AAT Immunology Trainee Abstract Award). Student have also been asked to speak at national conferences on their dissertation work.

7. Culture. The Department of Pharmacology and Physiology fosters a student-driven culture. Students in the department feel comfortable asking for advice from professors or fostering collaborations between labs. The students feel they chose their current mentor based not only on merit of research, but also on the departmental culture in Pharmacology and Physiology.

As a result of the meeting, Dean Wilmott announced on June 26, 2019 that Dr. Daniela Salvemini would be appointed interim chair of the Department of Pharmacology and Physiology. He also informed the department that a committee would be appointed to discuss the future of the department. I am happy to state that in March 2020, Dean Wilmott announced that a merger of the basic science departments would not take place. This means that the Department of Pharmacology and Physiology will remain intact and a search for a permanent chair will be conducted.

FIG. 44 Dr. Daniela Salvemini

Closing Comments

The Departments of Pharmacology, Physiology, and Pharmacology and Physiology at Saint Louis University School of Medicine have enjoyed a glorious, exciting, and productive history from the beginning in 1842 until the present time. To date there have been ten chairs of the Department of Pharmacology, with one acting chair; seven chairs of the Department of Physiology, with three acting chairs; and five chairs of the Department of Pharmacology and Physiology, with two acting/interim chairs. As far as is known, to date 152 individuals have served on the full-time faculty in these departments. It is quite possible that there have been some omissions in this number, especially those serving as instructors. If so, please accept the sincere apology of the author. Many of these faculty and chairs have achieved national and international fame as scientists, scholars, and educators. Over the years, faculty in these departments have generated millions of dollars in extramural support to fund their research, published thousands of peer-reviewed papers and other scholarly works, been invited to participate in numerous national and international meetings and conferences, given talks and seminars throughout the world, served on myriad study sections and review groups, participated on many editorial boards, and served on hundreds of local, national, and international committees. Many of these faculty have also distinguished themselves as chairs of academic departments, leaders, and deans of medical schools or other institutions of higher learning. In addition to the full-time faculty, numerous others have served with secondary, adjunct, or research track positions in the department.

Besides the hundreds of medical, nursing, dental, and allied health students educated in the department, 176 individuals as of 2019 have earned PhDs in the Departments of Pharmacology, Physiology, or Pharmacology-Physiology and another sixteen were mentored by faculty in the department, bringing the total to 193. Again, my apologies for any omissions. Numerous other individuals have been mentored as postdoctoral fellows or visiting faculty. Many of these former trainees have also gone on to distinguish themselves as scientists, scholars, and educators, or leaders in academia, industry, or government.

It is only fitting that this book is brought to a close in the year 2018 as the university celebrates the bicentennial of its founding. Saint Louis University can be proud of its Departments of Pharmacology, Physiology, and Pharmacology and Physiology in the School of Medicine over the years. The department is currently under the very able leadership of Dr. Daniela Salvemini as interim chair and the Department of Pharmacology and Physiology can look forward to many more years of outstanding contributions to and leadership at Saint Louis University.

References

1. Annual Catalogues and Announcements of the Medical Department of the St. Louis University, 1840–1855.
2. Adams, Rita Grace, William C. Einspanier, and B. T. Lukaszewski. *Saint Louis University: 150 Years*. Saint Louis University Libraries Digitization Center.
3. Faherty, William Barnaby. *Men to Remember: Jesuit Teachers at Saint Louis University, 1829–1979*. Saint Louis University Libraries Digitization Center.
4. Hill, Walter Henry. *Historical Sketch of the St. Louis University: The Celebration of Its Fiftieth Anniversary or Golden Jubilee, on June 24, 1879*. Saint Louis University Libraries Digitization Center.
5. Faherty, William Barnaby. *Better the Dream: Saint Louis, University & Community 1818–1968*. Saint Louis University Libraries Digitization Center.
6. *Bulletins of Saint Louis University: Announcements of the School of Medicine* (1904–present).
7. Franke, Florent E. *History of the Physiology Department of St. Louis University*. Unpublished, 1966.
8. Missouri Medical College Records/Bernard Becker Medical Library Archives; ID RG/R801H, Washington University School of Medicine in St. Louis.
9. "Life of Dr. William Beaumont, 1785–1853—Father of Gastric Physiology." Retrieved from https://www.james.com/beaumont/dr_life.htm
10. "William Beaumont, MD: The Prepared Mind." *Nutrition Today*, 6:28–32, 1971.
11. Wolf, S. "William Beaumont: The Man, His Time and His Legacy." *Federation Proceedings*, 44:2887–2888, 1985.
12. VanLiere, Edward J. "Alexis St. Martin, 1774–1880." *The Pharos*, 105–112, 1963.
13. Soule, S. D. "Dr. William Beaumont." *St. Louis Medicine*, 193–198, 1977.
14. Modlin, Irvin M. "From Prout to the Proton Pump—A History of the Science of Gastric Acid Secretion and the Surgery of Peptic Ulcer." *Surgery, Gynecology & Obstetrics*, 170:81–96, 1990.
15. Brodman, E. "William Beaumont and the Transfer of Biomedical Information." *Federation Proceedings*, 44:9–17, 1985.
16. Myer, Jesse S. *Life and Letters of Dr. William Beaumont, Including Hitherto Unpublished Data about the Case of Alexis St. Martin*. St. Louis, MO: C. V. Mosby, 1912.
17. Beaumont, William. *Experiments and Observations on the Gastric Juice, and the Physiology of Digestion*. Edinburgh, UK: MacLachlan & Stewart, 1838.
18. Garrison, F. H. *History of Medicine*, 4th ed. London, UK: W. B. Saunders, 1929.
19. St. Louis Medical and Surgical Journal History
19a. *Encyclopedia of the History of Missouri: A Compendium of History and Biography for Ready Reference*. Howard L. Conard, ed. New York, NY: The Southern History Company, 1901.
20. Schwitalla, Alphonse. Writings, Letters and Essays of Dr. Alphonse Schwitalla, S.J., Dean of the School of Medicine (1928–1948).
21. *The American Society for Pharmacology and Experimental Therapeutics, Incorporated: The First Sixty Years, 1908–1969*. K. K. Chen, ed. Washington, DC: The American Society for Pharmacology and Experimental Therapeutics, Inc., 1969.
22. Byrnes, Dolores with Padberg, J.W. and Waide, J. *Always the Frontier: Saint Louis University*, 1818-2018.